NATURAL DISASTER RESEARCH, PREDICTION AND MITIGATION

NATURAL DISASTERS

TYPHOONS AND LANDSLIDES - RISK PREDICTION, CRISIS MANAGEMENT AND ENVIRONMENTAL IMPACTS

NATURAL DISASTER RESEARCH, PREDICTION AND MITIGATION

Additional books in this series can be found on Nova's website under the Series tab.

Additional e-books in this series can be found on Nova's website under the e-book tab.

NATURAL DISASTER RESEARCH, PREDICTION AND MITIGATION

NATURAL DISASTERS

TYPHOONS AND LANDSLIDES - RISK PREDICTION, CRISIS MANAGEMENT AND ENVIRONMENTAL IMPACTS

KARINA LINWOOD
EDITOR

nova publishers
New York

For permission to use material from this book please contact us:
Telephone 631-231-7269; Fax 631-231-8175
Web Site: http://www.novapublishers.com

Additional color graphics may be available in the e-book version of this book.

Library of Congress Cataloging-in-Publication Data

Linwood, Karina.
Natural disasters : typhoons and landslides-- risk prediction, crisis management and environmental impacts / editors, Karina Linwood.
pages cm
Includes index.
ISBN 978-1-63463-309-3 (hardcover)
1. Natural disasters. 2. Natural disasters--Risk assessment. 3. Natural disasters--Environmental aspects. 4. Emergency management. 5. Typhoons. 6. Landslides. I. Title.
GB5014.L56 2014
551.55'2--dc23
2014039717

Published by Nova Science Publishers, Inc. † New York

CONTENTS

PREFACE

Typhoons pose a threat for coastal populations, offshore and onshore industries, shipping, agriculture, or forestry. Thus it is of great interest to assess future changes in cyclone activity. Thereby tropical cyclones may change both in frequency and intensity or they may have different track locations, shapes, dimensions or lifetimes. This book discusses risk prediction, crisis management and economic and environmental impacts of typhoons and landslides.

Chapter 1 – Typhoons pose a threat for costal populations, offshore and onshore industries, shipping, agriculture, or forestry. Future predictions of those phenomena are important, and depend to a great extent on the evaluation of changes in typhoon statistics of the past. In this study an analysis of typhoon frequencies and intensities over the last more than six decades was conducted. 'Best Track Data' sets were assessed, these are provided by several meteorological institutes, which compile reanalyzed in-situ- and satellite-based measurements of tropical cyclones (TC). But, the intensity estimations provided by individual data sets differ, which influences the derived climatological statistics. A source for these discrepancies are the varying approaches to estimate TC intensities used by the individual weather services, like the definition of TC intensity or measurement products. As a result, TC activity trends are ambiguous and depend on the analyzed data set.

An alternative approach to obtain typhoon statistics is to dynamically downscale reanalysis data with a regional climate model. Ensemble simulations are useful to evaluate the effect of internal model variability on typhoon simulations. A spectral nudging technique is beneficial for constraining the RCM's internal variability, as it keeps the large-scale weather state inside of the model domain close to the forcing reanalysis.

A long-term climatology was computed with the regional climate model CCLM and either the NCEP/NCAR or the ERA40 reanalyses as external forcing. The variability of intense TC numbers shows large similarity to the observed one, especially for the last three decades (when satellite data were available). But, long-term statistics derived for the whole simulation period (six decades) differ. Both RCM simulations indicate mostly an increase in TC numbers, while Best Track Data exhibit decadal variability. The RCM runs presumably reflect in homogeneities of the forcing reanalyses. An upward shift in TC intensities and in the magnitude of its variability in 1978 was detected in both regional simulations. At this time satellite measurements were introduced to the reanalyses. This seems to be responsible for the detected shift, and may contribute to the increase in TC activity measures.

Chapter 2 – The author's deal with simulation of great earthquakes accompanied with ensuing huge tsunamis in regions under typhoons. The main goal of this chapter is to present an idea of the proposal of the research project how to simulate the earthquake origins and how to simulate ensuing processes evoked by the great earthquakes like the tsunamis, submarine landslides and mudslides, that are in addition also under the presence of the typhoon, and thus to simulate how to preclude their consequences on the adjacent continental landscapes. Mathematical models will be presented and the corresponding mathematical methods and algorithms for their solutions will be introduced.

The presented methods will give some estimation about situations in the distressed areas closed the subduction zones and then after simulations of all the above mentioned processes before the real events they can give the necessary steps to preclude tragical situations in the large distressed continental areas.

Chapter 3 – Severe typhoon generates violent aerodynamic excitation on coastal cities and they can induce destructive effects on the civil infrastructures. This chapter aims at investigating the wind characteristics of severe typhoon and the corresponding wind-induced effects on reinforced concrete building. For this purpose, full-scale monitoring on a 64.70m height reinforced concrete building was conducted during the passage of severe typhoon. An ultrasonic anemometer and geophone-based accelerometers were installed on the building to capture continuous measurements including the wind velocity and structural response. The measurements are utilized to identify the wind characteristic properties and evaluate the wind-induced structural effects. The analysis is carried out by employing the Bayesian probabilistic framework. Exploiting the full advantages of Bayesian inference, not only the optimal values of the concerned parameters can be estimated but the associated estimation uncertainty can also be quantified in the form of probabilistic distribution. Consequently, abnormal changes of the estimates can be distinguished explicitly from the variation due to statistical noise. The result provides a reliable basis for the understanding of the features of severe typhoons and diagnosing possible structural damage.

Chapter 4 – Strong wind and heavy rain accompanied by typhoon passage induce physical disturbances such as upwelling, vertical mixing, terrestrial runoff and sediment resuspension, which supply nutrients to the photic layer in coastal waters. Consequently, phytoplankton biomass and production increase significantly. Primary production increased up to 19 times compared to that of pre-typhoon condition. Assimilation number (ratio of primary production to chlorophyll *a* concentration), which is an indicator of phytoplankton activity, is enhanced significantly after the passage of typhoon. In a temperate bay, a high assimilation number lasted for nine days after the passage of typhoon. The integrated primary production during the nine days after the passage of typhoon accounted for 7.2-9.1% of the annual primary production in the upper waters of the bay. Approximately three typhoons approach the bay annually, which implies that primary production in the upper waters of the bay may be enhanced in the range of 21.6-27.3% of the annual primary production.

One of the major fates of a phytoplankton bloom is the sinking of cells from the euphotic zone to deep waters. Typhoon-induced phytoplankton bloom is characterized by dominance of diatoms, such as *Skeletonema* spp., *Chaetoceros* spp., *Leptocylindrus* spp. and *Pseudo-nitzschia* spp. As diatoms show high sinking rates, their blooms may enhance the sinking flux of biogenic carbon to deep waters after the passage of typhoon. Another factor to consider is micro- and meso-zooplankton grazing on phytoplankton.

An abrupt decline in phytoplankton biomass and increase in zooplankton fecal pellets after the passage of typhoon suggests termination of the bloom due to intense zooplankton grazing.

In this chapter, the impact of typhoon on the phytoplankton biomass, production and species composition, and the fate of phytoplankton in coastal waters is documented and the ecological response of phytoplankton to the passage of typhoon is discussed.

Chapter 5 – The typhoon or hurricane is an atmospheric eddy originated in tropical or subtropical ocean regions. Because the typhoon center always moves, then the wind field becomes non-uniform and non-stationary. Hence, the typhoon wave estimation is more complicated than that of the monsoon. However, the typhoon swell propagates outwards in all directions and reaches the shore earlier. As the typhoon is approaching a site, the swell wave height will be enhanced due to the accumulation of wave energy flux which bears some resemblance to Doppler effect. There are many cases that freak waves cause accidents and damages while typhoons are far away. The assumption of the proposed typhoon swell prediction method is that the typhoon wave field is considered as a point source. The wave at the typhoon center is estimated by Bretschneider's parameter method. Empirical constants are obtained from measurements in Taiwan. The swell period will be longer, when the typhoon is more remote. The long wave is dangerous for the rock fishing, because the long wave can generate higher run-up and result to a larger backwash. However, the typhoon swell freak wave cannot be predicted but its potential can be alarmed.

Chapter 6 – A typhoon disaster database system incorporating information about typhoon track, weather, and typhoon-related disasters, which have occurred in Japan since 1951, was developed and opened to the general public, via the Internet, beginning in July 2005. To construct the system, information was collected mainly from the Japan Meteorological Agency, the Fire and Disaster Management Agency, and local governments. The system shows the spatial distribution of typhoon disasters on a map using WebGIS. In this chapter, the function of the system is introduced, and the relationship between the number of users accessing the system and the characteristics of typhoons from 2006 to 2012 are discussed. The peak number of unique visitors each day was independent of typhoon intensity. However, it increased when typhoons passed through the Tokai and Kanto regions, where the population is relatively high.

Chapter 7 – This study objectively investigates the potential of rainfall that could trigger the earthquake from the earthquake number (5.0 $\geqq$ML $\geqq$3.0) and the information of rain during the period from January 1995 to August 2012. (Lin 2013) reported the similar work that the typhoons could trigger small earthquakes easily especially after Chi-Chi earthquake during January 1995 to July 2012. The earthquakes are likely to have significant positive relationship with rainfall after the Chi-Chi earthquake with Chi-Square test. The result revealed a significant difference between the correlations of daily-accumulated rainfall values and earthquake activities both before and after the Chi-Chi earthquake. The significant difference was discussed regarding) changes of crustal conditions after the Chi-Chi earthquake.

Chapter 8 – Landslides process is one of the most widespread and dangerous processes in the urbanized territories. In Moscow the landslides occupy about 3 % of the most valuable territory of the city. There are near 15 places of deep landslides and some hundreds of shallow landslides in Moscow. In Russia many towns are located near rivers on high coastal sides. There are many churches and historical buildings on high costs of Volga River and Moscow River. The organization of monitoring is necessary for maintenance of normal functioning of city infrastructure in a coastal zone and duly realization of effective protective actions. Structure of monitoring system for urban territories is elaborated. Last years the landslide process activization took place in Moscow. One of landslide sites is on Vorob'ovy mountains, on a high slope of the right coast of the river Moscow with height of 65 m. In June 2007 rather big landslide took place there near ski-jump. Another landslide site is in a southeast part of Moscow, occupying the right coast of river Moscow near museum - reserve "Kolomenskoye". Last serious activization of a landslide has taken place there in 2002 with a motion on 53 cm. Catastrophic activization of the deep blockglide landslide in the area of Khoroshevo in Moscow took place in 2006-2007. Deep landslide activization took place in Moscvorech'e, Moscow, in September 2009. Such complicated situation demands development of new investigations of landslide prone zones.

Another part of activity was connected with Russia-Taiwan project "Theory and Methods of Earthquake Early Warning Systems for Underground Pipelines and Hazardous Slopes". Seismic monitoring post was established in Chelungpu fault area near Chiufengershan landslide for seismological phone measurements. Analysis of seismological phone changes will give possibility to elaborate early warning system. One of research methods for landslide processes is mechanical-mathematical modeling of gravitational movement of matter on landslide slope. At different stages of the process the development of landslide movement can be described by various mechanical and rheological models. Mechanical-mathematical model of high viscous fluid was used for modeling of matter behavior on landslide slopes. Equation of continuity and an approximated equation of the Navier-Stockes for slow motions in a thin layer were used. The results of modelling give possibility to define the place of highest velocity on landslide surface, which could be the best place for deformation monitoring post position. Model can be used for calibration of monitoring equipment and gives possibility to investigate some fundamental aspects of matter movement on landslide slope.

In: Natural Disasters
Editor: Karina Linwood
ISBN: 978-1-63463-309-3

Chapter 1

CHANGES IN TYPHOONS OVER THE LAST DECADES AS GIVEN IN OBSERVATIONS AND CLIMATE MODEL SIMULATIONS

Frauke Feser[1]* and Monika Barcikowska[2]
[1]Institute for Coastal Research, Helmholtz-Zentrum Geesthacht, Geesthacht, Germany
[2]Princeton Environmental Institute, Princeton University, Princeton, NJ, US

ABSTRACT

Typhoons pose a threat for costal populations, offshore and onshore industries, shipping, agriculture, or forestry. Future predictions of those phenomena are important, and depend to a great extent on the evaluation of changes in typhoon statistics of the past. In this study an analysis of typhoon frequencies and intensities over the last more than six decades was conducted. 'Best Track Data' sets were assessed, these are provided by several meteorological institutes, which compile reanalyzed in-situ- and satellite-based measurements of tropical cyclones (TC). But, the intensity estimations provided by individual data sets differ (Barcikowska et al., 2012), which influences the derived climatological statistics. A source for these discrepancies are the varying approaches to estimate TC intensities used by the individual weather services, like the definition of TC intensity or measurement products. As a result, TC activity trends are ambiguous and depend on the analyzed data set.

An alternative approach to obtain typhoon statistics is to dynamically downscale reanalysis data with a regional climate model (Feser and von Storch, 2008a, b). Ensemble simulations are useful to evaluate the effect of internal model variability on typhoon simulations (Feser and Barcikowska, 2012). A spectral nudging technique (von Storch et al., 2000) is beneficial for constraining the RCM's internal variability, as it keeps the large-scale weather state inside of the model domain close to the forcing reanalysis.

A long-term climatology was computed with the regional climate model CCLM and either the NCEP/NCAR or the ERA40 reanalyses as external forcing. The variability of intense TC numbers shows large similarity to the observed one, especially for the last three decades (when satellite data were available). But, long-term statistics derived for the whole simulation period (six decades) differ. Both RCM simulations indicate mostly

* Email: Frauke.Feser@hzg.de

an increase in TC numbers, while Best Track Data exhibit decadal variability. The RCM runs presumably reflect inhomogeneities of the forcing reanalyses. An upward shift in TC intensities and in the magnitude of its variability in 1978 was detected in both regional simulations. At this time satellite measurements were introduced to the reanalyses. This seems to be responsible for the detected shift, and may contribute to the increase in TC activity measures (Barcikowska et al., 2014).

1. Introduction

Tropical cyclones (TCs) can have devastating effects on coastal populations and their possessions, nature, agriculture and industry. Thus it is of great interest to assess future changes in cyclone activity. Thereby TCs may change both in frequency and intensity or they may have different track locations, shapes, dimensions or lifetimes. An indication for future alterations can be provided by changes of the past. To assess these reliably, very homogeneous time series of weather data are needed for long time scales of decades up to centuries. There are several ways to gain such hindcasts. First, reanalyses provide a realistic description of the past weather state, but those which are available for more than about 35 years (which is needed to compute reliable storm statistics) suffer from quite coarse resolutions of more than one hundred up to several hundred kilometers. For extreme weather events like TCs this grid distance is not narrow enough to simulate realistic low pressure cores and high wind speeds.

Best track data (BTD) describe sea level pressure and near-surface wind speeds along TC tracks. These start in the 1950s and therefore provide long-term data to compute storm statistics for the past decades. Webster et al., (2005) and Emanuel (2005) analyzed BTD sets and claimed there would be an increase in the number of the most intense TCs in the Western North Pacific. But according to Wu et al. (2006), who used several BTDs, neither the numbers of the most intense TCs nor the power dissipation index (PDI) defined by Emanuel (2005) show an increasing tendency. Ren et al. (2011) compared three BTD sets and confirmed increasing TC tendencies for one of them, but found decreasing for the other two. Kamahori et al. (2006) describe increasing numbers of TC days for weak TCs and decreasing numbers for stronger cyclones in one BTD set, while opposite trends were detected in another BTD set. These studies indicate a great dependency of the detected TC trends on the chosen BTD which may lead to different results.

Global climate models (GCMs) with small grid distances of up to 20 km have been used in a number of future scenario studies (e.g., Camargo and Barnston, 2009; Murakami et al., 2012a, b; Oouchi et al., 2006; Vitart et al., 1997). Many simulations return a decrease in TC frequency (Sugi et al., 2002; Sugi and Yoshimura 2004, 2012), but an increase in intensity. But these model runs cause high computational costs at this high resolution and do not depict past decades as observed. Therefore we here describe a dynamical downscaling approach, whereby global reanalysis data for the last more than 60 years is brought to a finer resolution, both in time and space, with a regional climate model (RCM). This strategy has been used in a number of applications, for Western Europe (Feser et al., 2001; Sotillo et al., 2005; Weisse et al., 2005) and also for Southeast Asia (Lee et al., 2004; Kang et al., 2005). Landman et al. (2005) showed that even with a relatively coarse resolution of 60 km, tropical storm tracks and vortex intensities can be simulated realistically by an RCM. Camargo et al. (2007)

analyzed the feasibility of downscaling a global model's seasonal TC activity with an RCM. They found that TC simulations were improved, even though their intensities did not reach observed values.

The purpose of downscaling is to resolve local weather patterns and features which are influenced by orography and regional meteorological processes. The forcing global model and the RCM have their particular best modelling performance at different spatial scales. Often a grid distance of five to six grid boxes is regarded to be sufficient to resolve weather phenomena. For global climate models this is in the range of about 500 km and larger, depending on the model's resolution. These well-resolved scales are considered to be trustworthy and should therefore not be altered by the RCM. Instead, the regional model should generate local meteorological features which were not provided by the coarse global model or reanalysis. But, for certain weather situations RCMs tend to differ from the observed large-scale state which is problematic for hindcast simulations which should stay close to the observations. This may happen when there are several atmospheric states which are equally likely to occur for the given conditions. Often this is the case for large RCM domains which allow for great internal variability and atmospheric conditions with only little exchange via the lateral boundaries (blocking weather situations). Castro et al. (2005) describe a reduced influence of lateral boundary conditions for the variability of small-scale features for larger model domains. Spectral nudging is a method to keep the RCM close to the forcing reanalysis for large weather phenomena while regional processes can be developed undisturbed. The method was applied by Kang et al. (2005) for an East Asian summer monsoon study. They describe an improved model performance for mean features as well as for single precipitation events.

In this chapter, first the quality of BTD is examined in terms of TCs of different strength and for BTD of several meteorological institutes. As an alternative, in the following a dynamical downscaling approach will be presented which regionalizes coarse-resolution reanalysis data for long time periods with a regional climate model. The ability of the model to simulate TCs realistically was evaluated and the role of spectral nudging in the formation of typhoons was assessed. In the last part TC variability on inter-annual to decadal time scales will be analyzed in comparison to observations. Finally, long-term TC activity changes are derived for the whole time span of more than six decades.

2. Tropical Cyclone Wind Speeds and Intensities in Best Track Data

Several data sets, so called Best Track Data, are provided by meteorological institutes in the U.S.A. (Joint Typhoon Warning Center (JTWC) and International Best Track Archive for Climate Stewardship (IBTrACS)), Japan (Japan Meteorological Agency (JMA)), and China (China Meteorological Administration (CMA)). BTD provide meteorological data, like core pressure or near-surface wind speed, along the tracks of TCs. In this chapter we will review the reliability of the different data sets for the Northwestern Pacific (NWP) and Southeast Asia. Barcikowska et al. (2012) examined TC tracks, lifetimes, and current intensity numbers for the NWP and showed that TC activity derived from the four BTD varies. Large differences can be seen for annually accumulated TC days separated into weak or strong categories according to the Saffir-Simpson-Hurricane Scale (SSHS). Similar trends are only found if all categories are analyzed as a whole.

A chief cause for the discrepancies in TC activity trends are different and non-stationary techniques used by the institutes to compute TC intensities. Examples are different averaging intervals of measured intensity. There are two approaches to minimize these differences, and their skill was evaluated. The first one is based on Atkinson's (1974) linear relationship between 10- and 1-min averaged intensities. Alternatively, Knapp and Kruk (2010) reversed intensity values back to operational, dimensionless parameters (current intensity numbers) and then applied a single conversion table to all datasets to achieve more homogenous intensity values. Both techniques are sufficient to reduce differences in TC activity trends and show an increase in annually accumulated TC days from 1978-2008, but only when TCs of SSHS categories 2 to 5 were taken into account as a whole. If categories 2 to 3 and 4 to 5 are regarded separately, the two techniques lead to different results.

Neither of the methods is sufficient to reconcile the differences among the most intense TCs (for categories 4 to 5). The linear multiplication factor overestimated TC day trends for lower categories, while the highest ones were still underestimated. The method of Knapp and Kruk (2010) worked fine especially for TCs of lower category (2 to 3), but removed the differences only in part for stronger TCs. So the converted BTD from JTWC still show a strong increase while converted JMA data (JMADT) experience no trend. In theory, the application of the same converting algorithm to the operational parameters to derive TC intensities for the different BTD sets should lead to similar results. But, some discrepancies still remain which indicate that additional factors contribute. One of them is that some weather agencies (e.g., JTWC) apparently used additional meteorological data irregularly in order to gain higher accuracy of TC center positions, which influenced the estimation of TC intensities. In contrast, JMA used mostly geostationary satellite data, which limits the accuracy. But this leads to larger homogeneity and thus makes the data set more useful to derive long-term statistics. Differences in current intensities correspond to deviations in TC center locations. Thereby largest discrepancies took place in the 1980s when satellite data was improving in resolution and availability. Later in the 1980s the reduction and phasing out of aircraft measurements may have had an influence. JTWC BTD show a strong increase in the number of intense TC days in the late 1980s until the end of the 1990s which coincides with track and current intensity differences between BTD sets.

To validate the BTD quality, scatterometer-based data were used as a reference for tropical storms, while aircraft measurements were used for higher intensities. For tropical storms JTWC shows systematically higher values than both JMADT and the reference data which leads us to the conclusion that JMA provides more realistic current intensities for these storms. For stronger storms sparse observation data limit a successful evaluation strategy. A comparison with aircraft measurements of 2008 and 2010 returned some agreement with maximum intensity estimations in BTDs. Unfortunately, no such data were available for earlier times when large discrepancies occurred.

The Dvorak technique (Dvorak 1972, 1973, 1975) estimates TC position and intensity from visible and infrared geostationary and polar orbiting weather satellite imageries. An analysis of strong typhoons like Sinlaku (2008) and Megi (2010) suggests that there are some deficiencies within the Dvorak technique procedures. BTD show slow weakening ratios in comparison to observations. We would like to emphasize how important documentation of the exact application and composition of the Dvorak technique is to interpret results. Also comparisons with aircraft measurements indicate that besides homogeneity a common strategy (harmonization and refinement of operational rules) for TC intensity estimations would be very desirable for all institutions that produce BTD sets.

3. A DYNAMICAL DOWNSCALING APPROACH

Due to the difficulties experienced with the Best Track Data an alternative approach was applied. For this, a regional climate model (RCM) was used to simulate the weather for Southeast Asia (SE Asia) and the Northwestern Pacific (NW Pacific) for the last more than six decades (Figure 1). We applied the RCM COSMO–CLM (CCLM, www.clm-community.eu; Rockel et al., 2008; Steppeler et al., 2003), which is the community model of the German regional climate research. It has been used for a wide range of applications at various resolutions up to cloud-resolving scales. For this study we used a spatial grid distance of 0.5° (about 53 km). The initial conditions, lateral and lower boundary forcing was provided by the global NCEP/NCAR reanalysis I (NCEP, Kalnay et al., 1996; Kistler et al., 2001) at a horizontal resolution of T62 (about 210 km). SSTs and sea ice information were taken from the NCEP reanalyses. Standard parameterizations were chosen. Test simulations were computed and resulted in a selection of the Tiedtke convection scheme (Tiedtke, 1989).

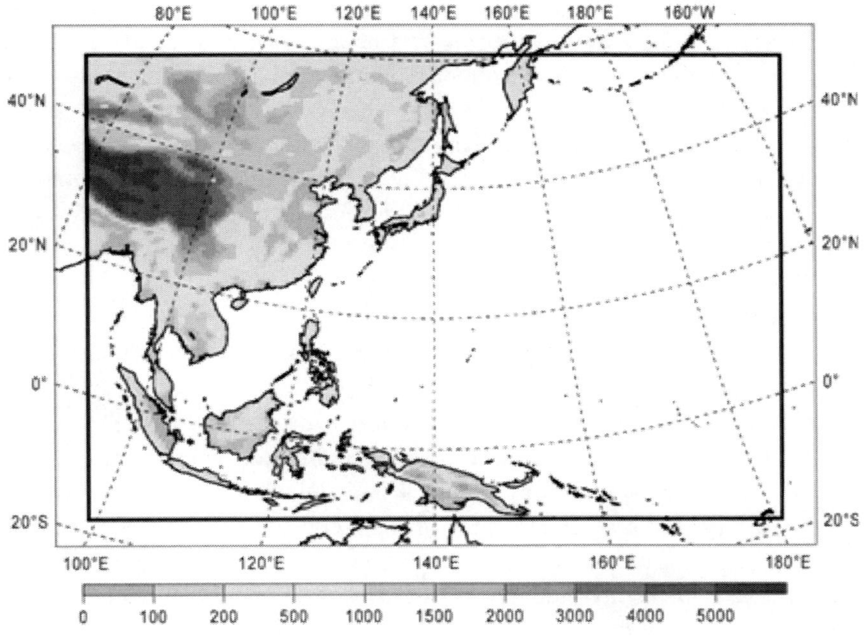

Figure 1. CCLM model area for Southeast Asia and the Northwestern Pacific. The grid distance is 0.5°. Grey scales denote surface elevation [m]. Figure redrawn after Feser and Barcikowska (2012), © IOP Publishing Ltd. CC BY-NC-SA.

In addition to the standard global forcing via the outer model boundaries, the spectral nudging technique (von Storch et al., 2000) is applied within the entire inner model domain. The method takes care that the large spatial scales of the regional model solution do not differ too much from the forcing global model or reanalysis. This is especially important for hindcast simulations of the past state which should be close to observations. First, the RCM results are transferred into spectral space, and then the large spatial scales are selected (low wave numbers; here it was applied for spatial scales larger than about 500-650km). Spectral nudging terms are added (for every third time step) which signs depend on the difference between the RCM and the reanalysis. This way the RCM solution is pulled towards the NCEP reanalysis at these spatial scales. Spectral nudging is only applied for large-scale atmospheric patterns, regional and local weather events are freely determined by the RCM. In this study, the method was used only for horizontal wind components above 850 hPa (so that regional processes close to the surface would not be disturbed) with increasing strength for higher model levels.

3.1. Typhoons Simulated in Regional Climate Models

For Western Europe a high-resolution hindcast approach was successfully applied by Feser et al., (2001). It consisted of dynamical downscaling of the NCEP reanalysis with an RCM using spectral nudging for the entire model domain. Now this downscaling technique was transferred to a region in SE Asia and the Northwestern Pacific. In this geographical area TCs are important weather features that have a large impact on population, nature, industry, or agriculture. To test the ability of the RCM to simulate TCs, ensemble simulations were computed with and without spectral nudging (Feser and von Storch, 2008a). Thereby the simulations were performed in climate mode so that they are independent of the initial conditions. The simulation area was chosen to be very large so that typhoons are generated inside of the model domain (without introduction of an artificial vortex) and do not move into it via the lateral boundaries. Several typhoon cases were simulated and compared to the NCEP reanalysis, which served as boundary conditions, in reference to BTD. Thereby the focus was put on TC numbers, tracks, core pressure, and near-surface wind speed along the tracks.

As a first example supertyphoon Winnie of August 1997 was analyzed, core pressure and wind speed were tracked in the RCM and NCEP and compared to BTD from JMA. In a double-nesting approach the NCEP reanalysis was first dynamically downscaled to about 50 km and then to 18 km Δx. TCs which were only coarsely described by NCEP were tracked in the RSM ensemble simulations and showed a considerably deeper core pressure and higher wind speeds. This was also the case for typhoon Winnie which developed inside of the model domain and showed a much deeper core pressure and considerably higher wind speeds than the forcing reanalysis. The five ensemble simulations (with and without spectral nudging) were started on consecutive days, otherwise the model settings were left unchanged. The purpose was to analyze the dependency of the simulations to their initial states. Without nudging, the ensemble revealed large variability between its ensemble members and many additional typhoons formed which were not observed. Also the location, shape, and length of typhoon tracks differed largely between the ensemble members without nudging and they were mostly very different in comparison to BTD (Figure 2). Spectral nudging helped to keep the TC tracks close to the observed ones. This lead to the conclusion, that lateral boundary conditions alone are not sufficient to determine dynamic developments inside of the model domain for such hindcast simulations. Instead, the same lateral boundary conditions can lead to different developments in the model domain for certain weather conditions. This conclusion depends on the size of the model domain, in smaller domains the lateral boundary control is more stringent (Castro et al., 2005).

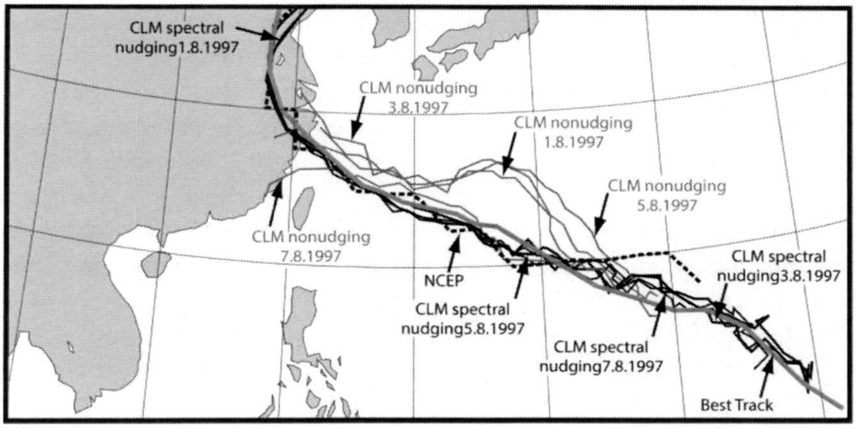

Figure 2. Tracks of typhoon Winnie as given by the JMA best-track data (thick grey solid line), by NCEP reanalyses (thick black dashed line), four simulations with CCLM without spectral nudging (thin grey lines), and four with spectral nudging (thin black lines). The only difference between the individual CCLM ensemble members with or without spectral nudging is their starting date at or shortly after 1 Aug 1997. Figure redrawn after Feser and von Storch (2008a).

In addition, several sensitivity experiments were conducted with different spectral nudging settings, like strength of the nudging coefficient or spatial scales to be nudged. But these changes in settings did not lead to large deviations from the former simulations with standard spectral nudging settings. The higher resolution (18 km) returned slightly improved core pressure and wind speeds along the tracks. The locations of typhoon tracks though were very similar between both resolutions.

In a subsequent study (Feser and von Storch, 2008b) the sample typhoon season 2004 was analyzed which featured many typhoons. The purpose was to validate the skill of the RCM in dynamical downscaling such extreme events from global reanalysis data and to test if it is feasible to simulate a high-resolution regional hindcast of weather in SE Asia and the NWP. Global NCEP reanalysis data were downscaled to about 50 km and in a double-nesting approach to 18 km Δx with the state-of-the-art RCM CCLM for this geographical region. Spectral nudging was applied inside of the model domain to keep the RCM close to the reanalysis for large weather phenomena which are simulated realistically in NCEP.

12 typhoons from the 2004 typhoon season were considered. These were examined in terms of sea level pressure (SLP) and near-surface wind speed in comparison to the forcing NCEP reanalysis and BTD from JMA as a reference. A tracking algorithm was used to find pressure minima and to select TC tracks and SLP and wind speeds along the tracks. For that certain criteria were applied like track length, duration, travel speed or wind speed in a distance around the pressure minimum. 10 of the 12 typhoons were detected in both the reanalysis and the RCM simulations. The comparisons revealed intensified SLP and wind speed values for the RCM which were mostly closer to BTD than the values of the global reanalysis. The RCM SLP along the track was often close to BTD, but for some TCs showed a lag in time. In most cases the double-nesting approach gave SLP slightly closer to BTD than

the coarser RCM simulation, for near-surface wind speeds it was the other way round (Figure 3). But still, both RCM resolutions were not sufficient to simulate intensities like in the BTD.

The distance of simulated tracks against observed tracks was measured by computing a great circle distance. It showed that though the RCM gave improvements in SLP and wind speed, NCEP always had the best representation of the track location. This is presumably a consequence of assimilation of TC data (including the exact location and track) in NCEP which is of course not available for the RCM which only gets information at the large spatial scales via the lateral boundaries and when spectral nudging is used also within the model domain. But the large spatial scales are not substituted, but only 'nudged' towards the NCEP reanalysis (and only at higher model levels above 850 hPa) which may lead to the slight dislocation of TC tracks. A validation for precipitation fields showed that the RCM can simulate the rain bands, rates, and patterns associated with extreme TCs much more realistically with higher resolution. Values were about twice as high for the 18 km simulation and showed values comparable to satellite data which were used as a reference.

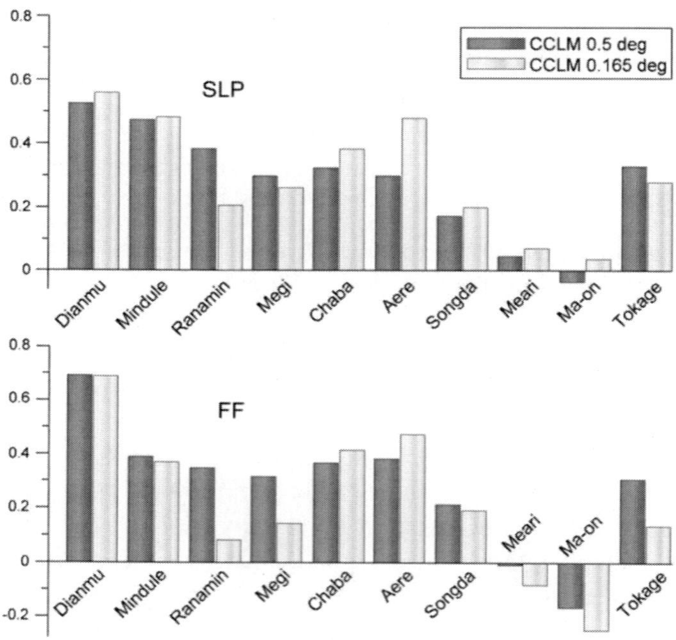

Figure 3. Brier Skill Score between JMA best track data and NCEP, CCLM 0.5°, and CCLM 0.165°, for SLP (upper panel) and 10 m wind speed (lower panel) for analyzed typhoons, for values larger than 0 CCLM is closer to the best track than NCEP, for 0 CCLM is equally close to the best track as NCEP, for values smaller than 0 NCEP is closer to BTD than CCLM. Figure redrawn after Feser and von Storch (2008b).

We conclude that strong TCs like typhoons can be depicted realistically by RCMs when global reanalysis data is dynamically downscaled. The RCM is able to simulate TCs inside of even large model domains without introducing an artificial vortex. Spectral nudging helps to keep the number of TCs as well as their location and tracks close to observations. The typhoon track is in most cases already located close to observations in the reanalysis, but improvements of the RCM can be expected for meteorological variables like, for instance,

SLP, near-surface wind speeds, or precipitation associated with the TC. But if some TCs are not represented by the forcing reanalysis, the RCM is not able to simulate these, as was the case for two typhoons which were not given by NCEP. Also a lag in time in the typhoon development cannot be corrected to a great extent. This shows the great dependency of the RCM on good quality of the forcing global model data.

3.2. Spectral Nudging for Better Typhoons Representations

Feser and Barcikowska (2012) analyzed the effect of spectral nudging (SN) on the simulation of TCs in regional climate simulations. Many studies (e.g., Cha et al., 2011; Feser and von Storch 2008a, b; Kanamitsu et al., 2010) showed before that RCMs are able to simulate TCs within large model domains without the introduction of artificial vortexes. One problem to derive long-term typhoon climatologies is the inherent internal model variability which may also affect the simulation of TCs and which can be quite large for spacious model domains. Spectral nudging has the capability to reduce the internal variability which leads to larger reproducibility of the model results and, on average, to larger similarity with observations (Weisse and Feser, 2003). This is especially important if only a single model simulation is feasible due to computational restrictions, like for long-term RCM simulations suited for the derivation of typhoon statistics. But the impact of the SN technique on TCs, associated large-scale circulations and weather patterns was yet not assessed.

In order to estimate the effect on the simulation of TCs, ensemble studies were computed with standard forcing via the lateral and lower boundaries or, in addition, with SN. First of all, the number of typhoons differed greatly between the simulations with and without SN. Thereby all of the ensemble simulations in standard configuration returned about twice as many TCs. But many of them had no counterpart in observations. With SN, the reduced number of TCs showed great similarity to BTD. This is depicted in Figure 4, which shows the distance of simulated TC track points compared to BTD for CCLM in standard configuration (left) and CCLM with SN (right). The size of the circles gives the number of track points which were classified into distance intervals of 100 km. The grey shades are different for each CCLM ensemble member. Three typhoons were not tracked in any of the RCM simulations. Five typhoons were only found for SN, two only for the standard runs. The figure shows a widely spread distribution for the differences of the simulations without SN, extending between 0–100 and 700–800 km. For the SN runs the numbers of tracks which are comparable to BTD tracks are much higher (bigger circles). The track differences converge towards smaller distance intervals of 100 and 200 km and become less frequent towards larger distance intervals.

Also pattern correlations for TC spatial densities and associated meteorological patterns were achieved. The standard simulations often showed TC locations and tracks which could not be seen in BTD. Those, which resembled the observations, deviated to some extent from the BTD tracks. For the SN ensemble simulations, the tracks converged soon after their first appearance and then stayed close to the BTD track. Meteorological fields which are important for the formation and development of TCs like temperature, precipitation and geopotential height showed larger similarity with observations for the SN ensemble. SN also led to a reduced ensemble spread which means that its members were more similar to each other which leads to a larger reproducibility of the results if only one simulation can be computed.

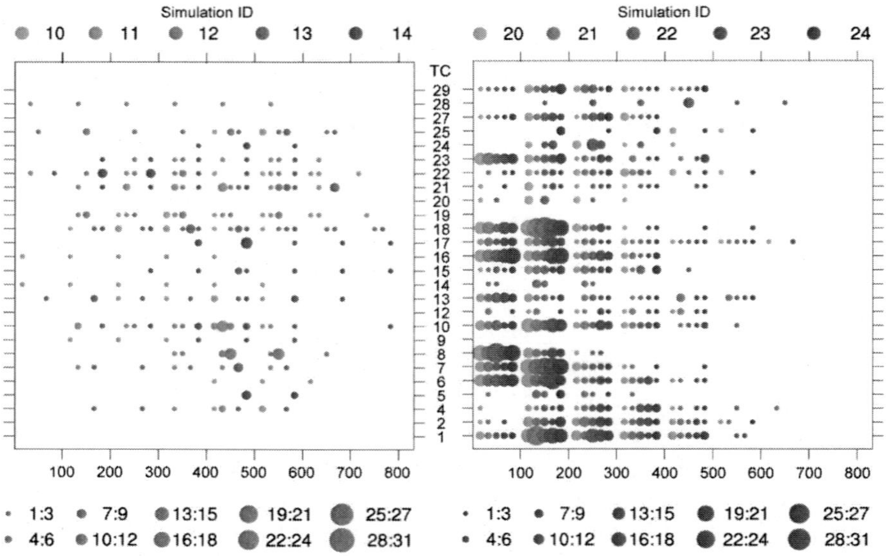

Figure 4. Difference of simulated TCs compared to BTD for the year 2004 for CCLM in standard configuration (left) and with SN (right). The y-axis shows the individual TC IDs, those TCs which were not tracked in either standard or SN simulations were omitted (3, 11, 26). The size of the circles on the x-axis gives the number of track points classified into distance intervals of 100 km. The different grey shades depict the individual CCLM ensemble members. Figure redrawn after Feser and Barcikowska (2012), © IOP Publishing Ltd. CC BY-NC-SA.

Cha et al., (2011) found reduced TC intensities and disturbed TC intensification processes caused by SN; therefore we looked into vertical temperature profiles for several typhoon cases. Here both ensembles showed comparable results which made us suggest that SN did not disturb the vertical development of the typhoons. General circulation patterns like monsoon or the Madden-Julian oscillation were not inhibited in the SN simulations; they were even more similar to a high-resolution reananalysis which was used as a reference. For wind speed distributions it was found that SN reduces the very extreme wind speeds a little in comparison to the standard configuration. But, we are still convinced that SN is essential for hindcast studies for SE Asia and the NWP, as it corrects the location, number, and track of TCs and thus leads to a realistic TC climatology for the last decades.

4. Long-Term Typhoon Climatologies

In order to obtain a TC climatology for the last decades, a dynamical downscaling approach can be applied to atmospheric reanalyses. Barcikowska and Feser (2014) analyzed TC climatologies derived with the regional climate model CCLM driven by the large-scale state prescribed by the NCEP/NCAR or the ERA40 reanalyses. A long-term TC climatology reconstructed with the CCLM model starting in 1948 until today shows good agreement with observations (BTD), but mostly for the recent three decades (Figure 7). For this period satellite measurements were available. They were used as a primary source facilitating forecasting practices and compilation of atmospheric reanalyses, e.g., BTD or NCEP-NCAR

Reanalysis. Consequentially, this period (1978-2011) was chosen to assess the skill of the downscaling methodology to reconstruct TC climatologies.

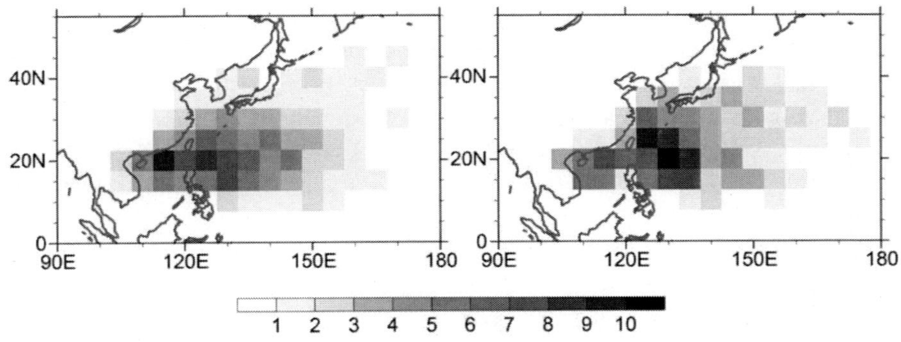

Figure 5. TC spatial densities (accumulated TC occurrence per every 5° latitude x 5°longitude grid box; normalized by mean spatial density fields) for a) BTD, b) CCLM in the period 1978-2011.

To verify the capability of CCLM to reproduce spatial features of TC tracks, reconstructed TC spatial density fields were compared to observations which were provided by JMA (Figure 5). The spatial density pattern of CCLM bears strong resemblance to the observed one, with a TC density maximum located between the Philippines, Taiwan and the China Sea. Leading EOFs (Figure 6) of annual reconstructed and observed TC fields depict trajectory patterns of intense TCs. The picture shows that CCLM reproduces precisely the observed pattern.

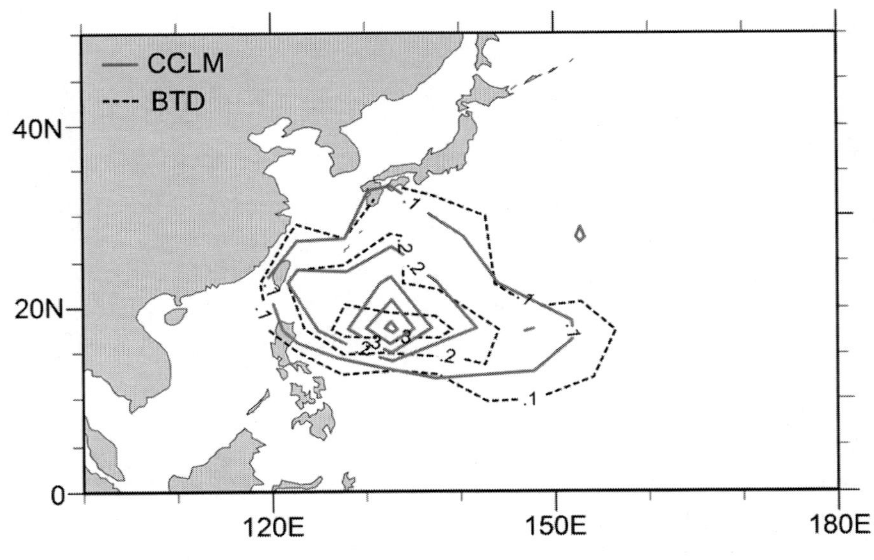

Figure 6. First EOF for the inter-annual variability of intense TCs (category 2-5 on the Saffir-Simpson Hurricane Scale) in CCLM simulation and BTD observations, for the common period 1978-2001. EOFs are shown for a) CCLM NCEP (grey solid line, total variance 31 %) and JMA (black stippled line, total variance 37 %).

The temporal variability of TCs during the last six decades was analyzed by means of the yearly number of TC days. Figure 7 shows strong inter-annual variability, both for reconstructed and observed TCs, which is caused to a great extent by the El Niño/Southern Oscillation (ENSO). The impact of ENSO on TC activity in the western North Pacific basin was already documented in many studies (e.g., Camargo et al., 2008). During the El Niño phase, the tropical Eastern and Central Pacific experience warm SST anomalies. These provide, together with positive air-sea interaction feedback, favorable conditions for TC genesis. As a consequence, TC genesis and trajectories over the western North Pacific are shifted eastward, leading to the development of higher intensities and longer lifetimes (e.g., annual number of TC days, as shown in Figure 7, or annually accumulated energy, Barcikowska and Feser, 2014).

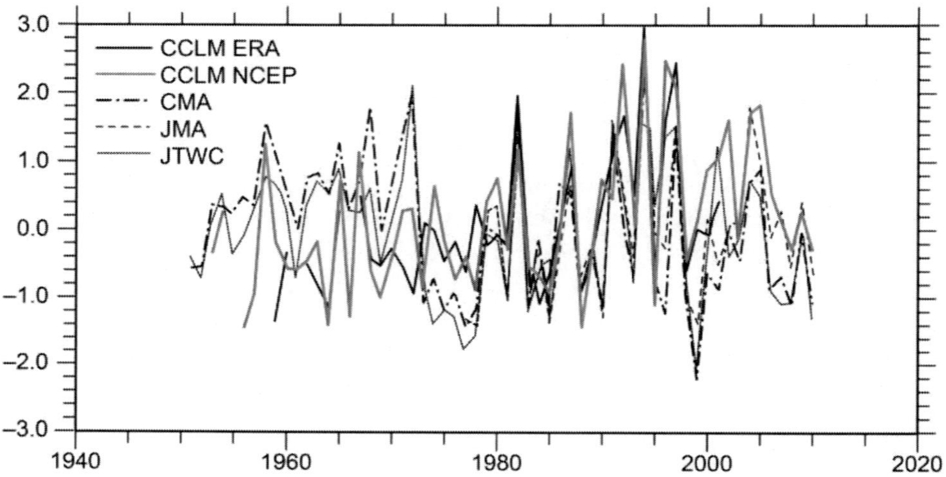

Figure 7. Annual TC days anomalies, normalized by standard deviation, and counted for all July-October seasons for CCLM simulations (forced with reanalyses: NCEP, ERA 40) and different BTD sets (JMA, JTWC, CMA) for intense TCs.

An analysis of inter-decadal variability in the recent three decades depicts similar features for observed and reconstructed data. These include overall upward tendencies and a remarkable drop since the late 1990s. The increasing trends may be long-term trends or part of multi-decadal variability (Chan and Liu 2004; Chan 2006; and Trenberth 2007) which represents, for instance, the low-frequency part of ENSO variability (Wang and Picaut, 2004).

Barcikowska and Feser (2014) show spatial features of the recent trends of CCLM NCEP, CCLM ERA, JTWC and JMA. All data sets indicate a north-eastward shift of TC activity over South-East Asia, which suggests an increasing contribution of TCs with longer, recurving trajectories. The magnitude of this feature varies among the data sets, but is most pronounced for CCLM ERA due to strong decreasing southward trends. In other data sets increasing TC activity tendencies prevail over the northwestern Pacific region. Overall, the downscaling method shows high skill in reproducing spatial and spectral TC climatology features. This is a motivation to analyze long-term TC reconstructions.

For the last six decades, which include the pre-satellite period, observations are inconclusive and indicate multi-decadal variability (for JTWC) or a decrease (for CMA).

Therefore, for this period the dynamical downscaling method is expected to provide alternative, more reliable information. In contrast to observations, the reconstructed TC variability by CCLM-NCEP (from 1948 to 2011) and CCLM-ERA (from 1951 to 1991) is dominated by a long-term upward trend (Figure 8). This difference obviously stems from the low agreement between observations and reconstructions before 1978, when the observational uncertainty was large. For this period, the reconstructed TC time series has much smaller values than observed. The reconstructed TC activity is also much smaller in these early years than for the latter period. This eventually contributes to the overall long-term increase.

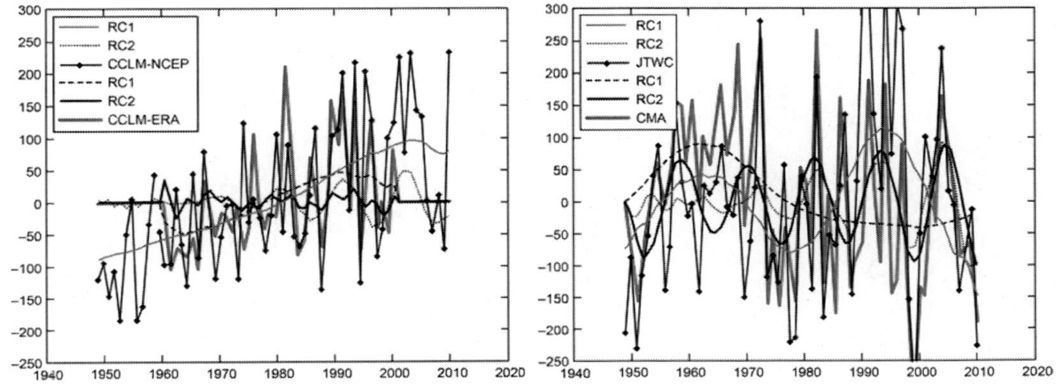

Figure 8. Decomposition of downscaled (a) and observed (b) TC variability using Multi-Channel Singular Spectrum Analysis, on the long-term component (RC1) and inter-decadal time scale variability (RC2). RC1 explains 32%, 23%, 19%, and 16% of the variability for CCLM-NCEP, CCLM-ERA, JTWC, and CMA respectively; while RC2 explains 8%, 9%, 10%, and 18%.

The reconstructed TCs exhibit inter-decadal variability. The CCLM-NCEP TC time series, shown in Figure 9, fits the PDO index well (r=0.33, with 1% significance level). It shows a similar increase from 1950 to the 1980s and a sharp decrease afterwards. Nevertheless, the PDO index cannot serve as an optimal reference to analyze low frequency features of TC data series. The standard EOF analysis is not designed to separate spectral features of the analyzed signal, thus the PDO index is contaminated largely by shorter frequency noise (e.g., interannual variability with an ENSO fingerprint) and its physical meaning is questionable.

Alternatively, Barcikowska and Feser 2014 applied a nonparametric and data-adaptive eigenvalue method (Multi-Channel Singular Spectrum Analysis, MCSSA) to select those frequencies that shape the analyzed TC variability. The MCSSA analysis was applied to the reconstructed and observed TC time series. It selected a pronounce inter-decadal signal (~15 yr) in all TC data sets. Figures 8a and b show a temporal evolution of the 15yr reconstructed component (RC1) extracted from all analyzed time series. Although the RC in the CCLM simulations is partly contaminated by shorter-term variability, its phase agrees remarkably well with the observed one. This temporal pattern was already found in SST records, with the largest contribution in the tropical Central and Eastern Pacific (Tourre et al., 2001). This again emphasizes the role of that region in shaping TC climatologies.

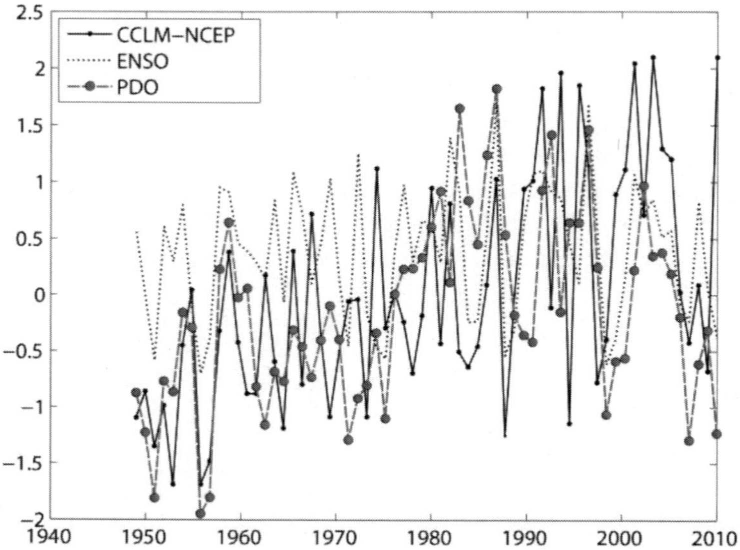

Figure 9. Annual TC days anomalies, standardized and counted for all July-October seasons for CCLM (NCEP-reanalysis-forced simulation, black solid line); Pacific Decadal Oscillation (PDO, stippled line with circles) index and the ENSO 3.4 index (dotted line).

The reconstructed TC intensity reflects upward tendencies found in the TC frequency. Figure 10 shows that nearly all of the TCs exceeding the 50th percentile of yearly maximum values occurred in the period after 1978. The difference between the pre-satellite period and the years after is even more striking for the leading EOF of reconstructed TC intensity fields (Figure 11). PC1, which represents a temporal evolution of the EOF mode, shows very low values and much smaller variance in the pre-satellite period.

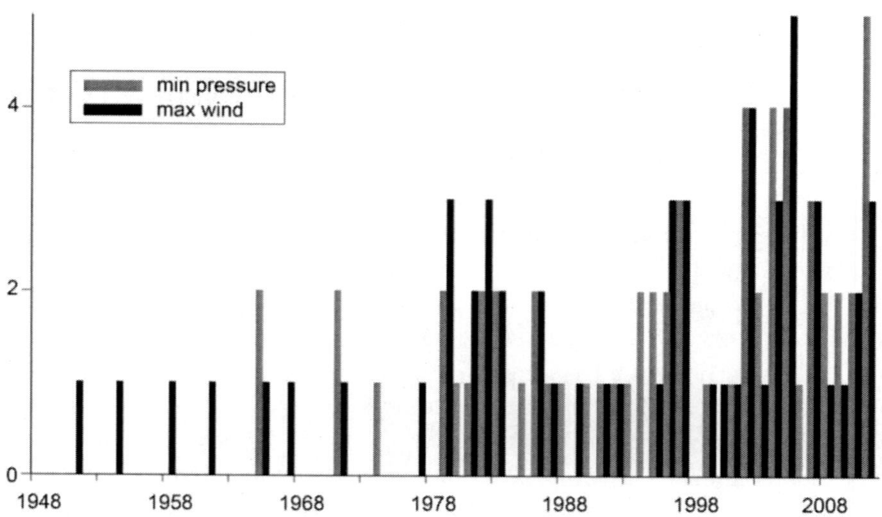

Figure 10. Number of TCs reaching an intensity threshold (50th percentile) of annual maximum of TC intensity.

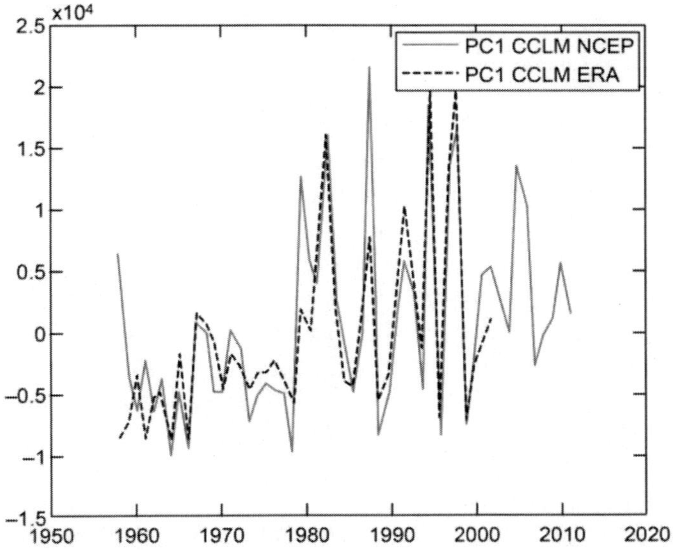

Figure 11. Time series of EOF pattern derived from TC intensity fields of CCLM NCEP (grey solid line) and CCLM ERA (black stippled line).

A remarkable shift towards positive values and increased variance since 1978 suggests that there are inhomogeneities in the Reanalyses, which impact the dynamically downscaled TC reconstructions. It is likely that the termination of the aircraft reconnaissance and the introduction of satellite measurements in 1978 impacted the quality and also the homogeneity of the Reanalyses. Therefore any interpretation of the results requires taking into account that introduced inhomogeneities may spuriously enhance the reconstructed upward TC activity trends.

CONCLUSION

BTD are computed by different meteorological services, they all use different techniques to derive TC intensities which lead to inconsistent TC long-term trends. In order to minimize these differences, several methods were applied. These are, for instance, a commonly used linear factor multiplication method to account for different averaging time intervals as well as the method of Knapp and Kruk (2010) which reconstructs TC intensity by remapping operational parameters (CI) with the same conversion table for all data sets. These methods show high skill to reduce trend discrepancies, but only when categories 2–5 are considered as a whole. But if less intense and strong TCs are considered separately, the techniques give varying results.

The application of the same converting procedures to retrieve TC intensities should, in theory, lead to the same intensities for the individual BTD. But the remaining differences indicate that there are additional contributing factors which cause the discrepancies. The largest differences happened in the 1980s when higher resolution satellite data became available. Also the reduction and termination of aircraft measurements presumably had an influence. BTD of JTWC show a strong increase in intense TC days between 1987 and 1999 which did not occur in other BTD. This may be a cause of adding new observational data and changing techniques over time. JMA uses mostly geostationary satellite data to estimate intensities which leads to limited data accuracy but also higher homogeneity which is essential to calculate statistics. The Dvorak technique is apparently applied differently by the meteorological services which produce BTD and a detailed documentation would be necessary. This is most important for extreme winds, where BTD sets show their greatest discrepancies.

Regional climate models may provide an alternative to BTD in the generation of long-term data sets suitable for typhoon statistics. Their advantage is the non-changing model system over time, with reanalysis or global model data as boundary and lateral conditions. The ability of RCMs to simulate TCs inside of their model domain was demonstrated also for large model domains without artificial vortexes. But in order to simulate typhoons at the right place and time for a hindcast simulation, some large-scale constraint like spectral nudging is necessary. The RCM can then achieve a deepened core pressure and higher near-surface wind speeds in comparison to coarse reanalysis data. The simulation of one typhoon season in 2004 showed that the RCM depicts the typhoons more similar to BTD, but failed to simulate those which were not included in the forcing global reanalysis data. Some SLP developments were close to BTD, but for others a time lag occurred which was already apparent in the reanalyses. We conclude that dynamically downscaled hindcasts depend greatly on the quality of the global input data. The RCM underestimated core pressure and wind intensities for all analyzed typhoons, which is presumably caused by the relatively coarse resolution of about 50 km. Higher RCM resolution (~ 18 km) lead to a slight improvement in terms of core pressure development and wind speed, but TC tracks were very similar to coarser simulations. The largest differences between both resolutions were found for precipitation. Rain bands around the TC centers were simulated much closer to satellite data, which served as a reference, when smaller grid distances were chosen.

The added value of regional climate modelling and the use of spectral nudging depend strongly on the type of application (Feser et al., 2011). If simulations of the recent past, for instance the last decades, are computed, the weather patterns and embedded extreme events like TCs should be simulated as realistic as possible. This is most important for the generation of homogeneous multi-decadal climatologies which shall serve as a source for statistical analysis of changes in TCs over time. Therefore we recommend using spectral nudging, since it leads to more realistic TC distributions, locations and tracks. Associated weather patterns show a larger similarity with reference high-resolution reanalyses data. An improvement in atmospheric circulations was found for monsoonal circulations and the Madden-Julian oscillation.

Overall, the dynamical downscaling approach, using atmospheric reanalyses (NCEP-NCAR and ERA 40) as input for an RCM, shows high skill in reproducing TC climatologies. For the period with the best measurement quality and availability (1978-2011), spatial and temporal features of the reconstruction bear close resemblance to the observed ones. Both reconstructed and observation-based data sets show an increase of TC activity, in terms of annually accumulated TC days for the last three decades.

If the approach is expanded towards earlier periods, when atmospheric measurements were scarce and less reliable, the results contest the conclusions derived from observation-based TC data sets. The reconstructed long-term climatologies, driven by both atmospheric reanalyses (for NCEP-NCAR in 1948-2011, for ERA 40 in 1951-1991), suggest a strong increase of TC activity, both in terms of annual TC days and TC intensity. In contrast, observational BTD sets are rather inconclusive and show decadal fluctuations for JTWC and a decrease for JMA. But it is possible that improving quality and availability of atmospheric measurements over time lead to inhomogeneities in both reanalyses. Therefore it is necessary to take into account that the strong increasing trends of the downscaled TC activity may partly be an artefact of changing data quality of the atmospheric fields driving the reconstructions.

ACKNOWLEDGMENTS

The work was partly supported through the Cluster of Excellence 'CliSAP' (EXC177), Universität Hamburg, funded through the German Research Foundation (DFG). It is a contribution to the Helmholtz Climate Initiative REKLIM (Regional Climate Change), a joint research project of the Helmholtz Association of German research centres (HGF). The model simulations were performed at the German Climate Computing Center (DKRZ). We would like to thank B. Gardeike for her help with the figures.

REFERENCES

Atkinson, G. D., 1974: Investigation of gust factors in tropical cyclones. *FLEWEACEN Tech. Note JTWC 74-1*, Fleet Weather Center, Guam, 9 pp.

Barcikowska, M. and F. Feser, 2014: Changes in tropical cyclone activity for the Western North Pacific during the last decades, derived from a regional climate model simulations, *Climate Dynamics*, submitted.

Barcikowska, M., F. Feser, and H. von Storch, 2012: Usability of best track data in climate statistics in the western North Pacific, *Mon. Weather Rev.*, 140, pp 2818-2830, doi: 10.1175/MWR-D-11-00175.1.

Camargo, S. J. and A. G. Barnston, 2009: Experimental dynamical seasonal forecasts of tropical cyclone activity at IRI. *Wea. Forecasting*, 24, 472-491. doi: http://dx.doi.org/10.1175/2008WAF2007099.1

Camargo, S. J., A. W. Robertson, A. G. Barnston and M. Ghil, 2008: Clustering of eastern North Pacific tropical cyclone tracks: ENSO and MJO effects. *Geochemistry, Geophysics and Geosystems*, 9, Q06V05.

Camargo, S., H. Li, and L. Sun, 2007: Feasibility study for downscaling seasonal tropical cyclone activity using the NCEP regional spectral model. *Int. J. Climatol.*, 27, 311-325.

Castro, C. L., R. A. Pielke Sr., and G. Leoncini, 2005: Dynamical downscaling: Assessment of value retained and added using the Regional Atmospheric Modeling System (RAMS). J. *Geophys. Res.*, 110, D05108, doi:10.1029/2004JD004721.

Cha, D.-H., C.-S. Jin, D.-K. Lee and Y.–H. Kuo, 2011: Impact of intermittent spectral nudging on regional climate simulation using Weather Research and Forecasting model. *J. Geophys. Res.*, 116, D10103.

Chan, J. C. L., 2006: Comments on “Changes in Tropical Cyclone Number, Duration, and Intensity in a Warming Environment”. *Science*, 311, 1713.

——, and K. S. Liu, 2004: Global warming and western North Pacific typhoon activity from an observational perspective. *J. Climate*, 17, 4590–4602.

Dvorak, V. F., 1972: A technique for the analysis and forecasting for tropical cyclone intensities from satellite pictures. *NOAA Tech. Memo*. NESS 36, NOAA, 15 pp.

——, 1973: A technique for the analysis and forecasting for tropical cyclone intensities from satellite pictures. *NOAA Tech. Memo*. NESS 45, NOAA, 19 pp.

——, 1975: Tropical cyclone intensity analysis and forecasting from satellite imagery. *Mon. Wea. Rev.*, 103, 420-430.

Emanuel, K., 2005: Increasing destructiveness of tropical cyclones over the past 30 years. *Nature*, 436, 686-688.

Feser, F. and M. Barcikowska, 2012: The Influence of Spectral Nudging on Typhoon Formation in Regional Climate Models. Environ. Res. Lett., 7, 014024, doi:10.1088/1748-9326/7/1/014024.

Feser F., B. Rockel, H. von Storch, J. Winterfeldt and M. Zahn, 2011: Regional climate models add value to global model data: a review and selected examples. *Bull. Am. Meteorol. Soc*., 92, 1181-92.

Feser, F. and H. von Storch, 2008b: Regional modelling of the western Pacific typhoon season 2004. *Meteorolog. Z.*, 17 (4), 519-528.

Feser, F. and H. von Storch, 2008a: A dynamical downscaling case study for typhoons in SE Asia using a regional climate model. *Mon. Weather Rev*., 136 (5), 1806-1815.

Feser, F., R. Weisse, and H. von Storch, 2001: Multi-decadal atmospheric modeling for Europe yields multi-purpose data. *Eos*, Trans. Amer. Geophys. Union, 82, 305-310.

Kalnay, E., M. Kanamitsu, R. Kistler, W. Collins, D. Deaven, L. Gandin, M. Iredell, S. Saha, G. White, J. Woollen, Y. Zhu, A. Leetman, R. Reynolds, M. Chelliah, W. Ebisuzaki, W. Higgins, J. Janowiak, K. C. Mo, C. Ropelewski, J. Wang, R. Jenne, D. Joseph, 1996: The NCEP/NCAR 40-year reanalysis project. *Bull. Am. Meteorol. Soc*. 77, 437-471.

Kamahori, H., N. Yamazaki, N. Mannoji, and K. Takahashi, 2006: Variability in intense tropical cyclone days in the western North Pacific. *SOLA*, 2, 104-107.

Kanamitsu, M., K. Yoshimura, Y.-B. Yhang and S.-Y.Hong, 2010: Errors of interannual variability and trend in dynamical downscaling of reanalysis. *J. Geophys. Res*., 115, D17115.

Kang, H. S., D. H. Cha and D. K. Lee, 2005: Evaluation of the mesoscale model/land surface model (MM5/LSM) coupled model for East Asian summer monsoon simulations. *J. Geophys. Res*., 110, D10105.

Kistler, R., W. Collins, S. Saha, G. White, J. Woollen, E. Kalnay, M. Chelliah, W. Ebisuzaki, M. Kanamitsu, V. Kousky, H. van den Dool, R. Jenne, M. Fiorino, 2001: The NCEP–NCAR 50-year reanalysis:Monthly means CD-ROM and documentation. *Bull. Am. Meteorol. Soc*. 82, 247-267.

Knapp, K. R., and M. C. Kruk, 2010: Quantifying interagency differences in tropical cyclone best-track wind speed estimations. *Mon. Wea. Rev*., 138, 1459-1473.

Landman, W. A., A. Seth, and S. J. Camargo, 2005: The effect of regional climate model domain choice on the simulation of tropical cyclone-like vortices in the southwestern Indian Ocean. *J. Climate*, 18, 1263-1274.

Lee D., K., D. H. Cha and H. S. Kang, 2004: Regional climate simulation of the 1998 summer flood over East Asia. *J. Meteorol. Soc. Japan*, 82, 1735-53.

Murakami, H., Y. Wang, H. Yoshimura, R. Mizuta, M. Sugi, E. Shindo, Y. Adachi, S. Yukimoto, M. Hosaka, S. Kusunoki, T. Ose, and A. Kitoh, 2012a: Future Changes in Tropical Cyclone Activity Projected by the New High-Resolution MRI-AGCM*. *J. Climate*, 25, 3237–3260. doi: http://dx.doi.org/10.1175/JCLI-D-11-00415.1.

Murakami, H., R. Mizuta, and E. Shindo, 2012b: Future changes in tropical cyclone activity projected by multi-physics and multi-sst ensemble experiments using the 60-km-mesh MRI-AGCM. *Climate Dynamics*, 39 (9-10), 2569{2584, doi:10.1007/s00382-011-1223-x.

Oouchi, K., J. Yoshimura, H. Yoshimura, R. MizutaI, S. Kusunoki, and A. Noda, 2006: Tropical cyclone climatology in a global-warming climate as simulated in a 20 km-mesh global atmospheric model: Frequency and wind intensity analyses. *Journal of the Meteorological Society of Japan*, 84, No. 2, 259-276.

Ren, F., J. Liang, G. Wu, W. Dong, and X. Yang, 2011: Reliability analysis of climate change of tropical cyclone activity over the western North Pacific. *J. Climate*, 24, 5887-5898.

Rockel B. W., A. Hense, 2008: The regional climate model COSMO-CLM (CCLM). *Meteorol. Z.* 12, 347-348.

Sotillo, M. G., A. W. Ratsimandresy, J. C. Carretero, A. Bentamy, F. Valero, and F. González-Rouco, 2005: A high-resolution 44-year atmospheric hindcast for the Mediterranean Basin: Contribution to the regional improvement of global reanalysis. *Climate Dyn.*, 25, 219-236.

Steppeler J., G. Doms, U. Schättler, H. W. Bitzer, A. Gassmann, U. Damrath, G. Gregoric, 2003: Meso-gamma scale forecasts using the nonhydrostatic model LM. *Meteorol. Atmos. Phys.* 82, 75-96.

Tiedtke, M., 1989: A comprehensive mass flux scheme for cumulus parameterization in large scale models. *Mon. Weather Rev.*, 117, 1779-800.

Tourre et al, 2001: Patterns of coherent decadal and interdecadal climate signals in the Pacific basin during the 20th Century, *Geophys. Res. Lett.*, 28, 2069-2072.

Trenberth, K. E., and Coauthors, 2007: Observations: Surface and atmospheric climate change. *Climate Change 2007: The Physical Science Basis*, S. Solomon.

Vitart, F., J. L. Anderson, and W. F. Stern, 1997: Simulation of interannual variability of tropical storm frequency in an ensemble of GCM integrations. *J. Climate*, 10 (4), 745-760, doi:10.1175/1520-0442(1997)010h0745:SOIVOTi2.0.CO;2.

von Storch, H., H. Langenberg, and F. Feser, 2000: A Spectral Nudging Technique for Dynamical Downscaling Purposes. *Mon. Weather Rev.*, 128 (10) 3664-3673.

Wang, C., and J. Picaut, 2004: Understanding ENSO physics — A review. *Earth's Climate: The Ocean–Atmosphere Interaction*, Geophys. Monogr., 147, Amer. Geophys. Union, 21-48.

Webster, P. J., G. J. Holland, J. A. Curry, and H. R. Chang, 2005: Changes in tropical cyclone number, duration, and intensity in a warming environment. *Science*, 309, 1844-1846, doi:10.1126/science.1116448.

Weisse, R. and F. Feser, 2003: Evaluation of a method to reduce uncertainty in wind hindcasts performed with regional atmosphere models. Coastal Engineering, 48 (4), 211-225.

Weisse, R., H. von Storch, and F. Feser, 2005: Northeast Atlantic and North Sea storminess as simulated by a regional climate model during 1958–2001 and comparison with observations. *J. Climate*, 18, 465-479.

Wu, M. C., K. H. Yeung, and W. L. Chang, 2006: Trends in western North Pacific tropical cyclone intensity. *Eos, Trans. Amer. Geophys. Union*, 87 (48), 537-538.

In: Natural Disasters
Editor: Karina Linwood
ISBN: 978-1-63463-309-3

Chapter 2

Mathematical Simulation of Earthquakes in Regions Under Typhoons: Possible Methods for the Risk Prediction; Mathematical Models, Computational Methods and Algorithms

Jiří Nedoma*
Institute of Computer Science, Academy of Sciences of the Czech Republic, Prague, Czech Republic

Abstract

We deal with simulation of great earthquakes accompanied with ensuing huge tsunamis in regions under typhoons. The main goal of this chapter is to present an idea of the proposal of the research project how to simulate the earthquake origins and how to simulate ensuing processes evoked by the great earthquakes like the tsunamis, submarine landslides and mudslides, that are in addition also under the presence of the typhoon, and thus to simulate how to preclude their consequences on the adjacent continental landscapes. Mathematical models will be presented and the corresponding mathematical methods and algorithms for their solutions will be introduced.

The presented methods will give some estimation about situations in the distressed areas closed the subduction zones and then after simulations of all the above mentioned processes before the real events they can give the necessary steps to preclude tragical situations in the large distressed continental areas.

Keywords: Earthquakes, tsunamiearthquakes, typhoons, tsunamis, landslides, mudslides, mathematical models, variational formulations of corresponding model problems, computational methods and algorithms

*E-mail address: nedoma@cs.cas.cz

1. Simulations of Consequences of the Tsunamigenic Earthquakes in Regions Under the Typhoons

1.1. Introduction

Regions with great climatic events such as typhoons and super-typhoons and with very active seismicity, represented with frequent seismic events and great earthquakes, both accompanied with ensuing huge tsunamis that evoke great deluges and landslides, are observed in several areas of the Earth. These events have been responsible for some of most destructive natural disasters in terms of human and economical loses. Therefore, the main aim of the part of the global research projects for such regions is to give an idea how to predict the future situations in these endangered regions and how to predict such future situations before the future events, and therefore, to predict also their future probable consequences. It is evident that both types of events - the typhoon and the tsunamiearthquake(s), together with the submarine landslides and the volcanic activity, that can operate at the same time and that can evoke the huge tsunamis with catastrophic consequences in these regions, such as reshaping the geology near the coast of the ocean by eroding sand from the large shore areas and destruction of a great part of the near-closed countries, represent one of the most detrimental natural hazards. The large parts of landscapes are reached by enormous quantities of water, that are accompanied by great deluges, and because the rocks are saturated by a water, landslides of soils and rocks can be also arised. Moreover, these regions are further handicapped by the effects of ensuing aftershocks and their immediate consequences. Therefore, the prediction of landscapes intervened with potential landslides and mudflows are necessary. Such possibilities allow to combine the data obtained from the warning systems (the typhoon warning system, the earthquake warning system, the tsunami warning system and the volcanic warning system) together with mathematical simulations of all processes taking place in these above mentioned regions, based on the mathematical simulation of tectonic motions and fractures of lithospheric plates and blocks with ensuing propagation of seismic waves, on the mathematical simulation of coupled thermo-hydro-mechanic and thermo-hydro-dynamic processes taking place in these regions, as the knowledge of typhoon's propagation and of the stress-strain states inside the upper parts of the lithosphere are fundamental for understanding of the future security in these mentioned regions. Moreover, such studies are also of a great importance for the design of large structures like the nuclear power plants, reservoir dams, etc. The idea of the proposal how to predict the future situations in the endangered region under typhoon is based on the mathematical simulaton of coupled thermo-mechanical and thermo-hydro-dynamical processes together with the oceanic and climatic processes in the endangered region during huge tsunami, typhoons and their ensuing consequences of enormous quantities of water in the endangered regions onto the stability of slopes in these regions and in general onto the security of these regions. The mathematical models will be based on the dynamic multi-body contact mechanics in the different rheologies. In our study the mathematical models will be based on the Bingham thermo-visco-plastic rheology and on the nonlinear thermo-visco-elasticity in the multi-body contact mechanics. The presented idea of the proposal allows to simulate the situation before the typhoon's and earthquake's events and thus it allows to test the safety arrangements in time, much more earlier before the real situations.

1.2. Tsunamigenic Earthquakes, Tsunamis, Typhoons and Their Detections

1.2.1. Geodynamic Processes in the Lithosphere and the Earthquake's Origins

The evolution of the lithosphere depends on the geodynamic processes inside the Earth. The Earth's lithosphere is divided into 14 large plates, 38 small plates and small blocks (Bird (2003), DeMets et al. (2010)), that mutually move. With opening the rift the oceanic lithosphere start to be originated. The newly formed oceanic lithosphere moves away from a mid-oceanic ridges (the divergent boundaries), its thickness increases and we speak about the sea-floor spreading. Its average thickness is less of about $80\,km$. In the oceanic lithosphere the transform faults are well defined by the fracture zones and are characterized by the absence of sedimentary covers. Transform faults, like the ocean ridges, are characterized by shallow earthquakes, that are localized in depths less than $50\,km$. These earthquakes are relatively small in magnitude, commonly occur in swarms and are accompanied also with volcanism. The ocean-ridge system is the longest topographic feature on the surface of the Earth and is of about $70000\,km$ in length. The zone between two plates sliding horizontally past one another is called the transform boundary and it is graphically depicted by '$\rightleftharpoons$". Transform faults may result structural discontinuities of large sizes on the bottom of the oceans and seas, and moreover, may result structural and topographical breaks - the so-called fracture zones. There are three types of transform faults, that is, ridge-ridge, ridge-trench, and trench-trench transform faults. The transform faults are occured also in continents as the continental transform faults and accompanied with the large earthquakes $(M > 8)$, separated by relatively long periods of quiescence of about $150\,yrs$ (i.e., on the San Andreas fault in California). Transform faults may arise at the junctions of ridge segments, that are necessary results of rigid plates on a spherical Earth's body. McKenzie and Morgan (1969) indicated sixteen possible combinations of ridge, trench, and transform-fault triple junctions (that is, points where three plates meet). But only six of them are common.

Distributions of the epicenters of earthquakes with large magnitudes indicate that their locations define plate boundaries. Earthquakes are characterized by their focal depths as earthquakes with shallow focus ($0-70\,km$ of depths), with intermediate focus ($70-300\,km$ of depths) and with deep focus (greater than $300\,km$ of depths).

Continental rifts are evoked by the existence of horizontal deviatoric tensional stresses, that are sufficient to break the continental lithosphere, while the evolution of the oceanic lithosphere was originated at the ocean rifts, ensuing seafloor spreading and cooling of rocks. The evolution of the continental lithosphere had very complicated evolution during the geological epochs. Because the geodynamic processes in the continental lithosphere are not of our interest, we do not analyzed them in this chapter.

Manifestations of underthrusting oceanic lithosphere is the oceanic trenches. There exist large regions of continental and oceanic nature where the lithosphere undergoes large-scale deformations and the diffuse seismicity prevails. Then we speak about subduction and collision zones. These zones are localities of fundamental geodynamic, geophysical and geological processes, that is, they are probably results of major plate-driving forces. The lithospheric plates move and then the weakened (thinner) plate descend below the thickened plate (the overriden lithospheric plate) and we speak about **subduction**. Subductions occur when the lithospheric plates move by velocities more than $50\,mm/yr$. Subduction is accompanied by development of deep-sea basins. A large number of events occur in the seismic zone - the Wadati-Benioff zone. This zone is characterized by brittle rocks of descending lithospheric slabs. The majority of the Wadati-Benioff zones are weakly curved, $10-30\,km$ thick, and they reach to depth of as much as $650\,km$, and moreover, their geometry is more or less curved to $200-300\,km$ depths. The Wadati-Benioff zones are also divided by aseismic zones into two or more seismically active zones. Bellow these depths occur more irregularities associated with aseismic zones. The double seismic zones beneath the Japan arc system are also observed (Hasegawa et al. (1978), Peacock (2001)). Seismic zones vary in hypocentre distribution as well as in dip, e.g. in the Aleutian arc system up to $300\,km$ of depths, in the Andean regions up to $> 600\,km$ of depths and in some of regions like the Northern Mariana arc system up to $700\,km$ of depths (Berrocal and Fernandéz (1996a,b), Condie (1997), (2003), Nedoma (1998), (2007), Berrocal et al. (2000), Karato et al. (2001), Kearay et al. (2009)). Oceanic trenches are results of underthrusting oceanic lithosphere, and are formed in places where the lithospheric slabs begin descend into the lower parts of the mantle. Subduction zones are potential tsunami locations.

The movements of lithospheric plates are evoked by the continental drifts. The rates of movements are called the rates of convergence of lithospheric plates and blocks. The oceanic arc systems are originated in places where the oceanic lithospheric plate is subducted (descended) beneath the other oceanic lithospheric plate. The island arc systems are characterized by the seismic and volcanic activities as well as the fore-arc and back-arc basins.

The ocean-continent convergence in the next stage in some regions leads to continent - continent convergence (e.g. the Himalayan-Alpine system). If both colliding lithospheric plates are of the same thickness the geodynamic process leads to evolution of collision zones.

The rates of subductions and of collisions of lithospheric plates are called the rate of convergences. In the case of subduction the lithospheric plates move faster, $> 50\,mm/yr$, and in the case of collision the lithospheric plates move more slowly. Then we distinguish two main types of orogens: (i) the Andean (or Cordillerian or non-collisional) type of orogens that originate from the ocean-continent convergence when the plate motions (more than $50\,mm/yr$) and other factors controlling subduction lead to compression within the overriding plates, and (ii) the collisional type of orogens that originate when the continental plates or the island arcs collide with a continental margin(s) as a result of the previous subduction.

Earthquakes that generate tsunamis are called tsunamigenic or the tsunamiearthquakes (Satake and Tanioka (1999)). The tsunamiearthquakes are generated in the subduction zones and are classified with respect to their location and faulting mechanism. The meaningful measure of the earthquake size is the seismic moment M_0 (Lay and Wallace (1995)), defined as

$$M_0 = \mu \cdot S \cdot d,$$

where S is the area of the fault, μ is the shear modulus, that is, the resistance of rocks to shearing, and d is the amount of the slip. The larger areas of faults band slip then the greater possibilities for the tsunamigenic potential of an earthquake exist. The other quantities for tsunamiearthquake determinations are the moment magnitude M_W defined as

$$M_W = \frac{2}{3} log_{10}(M_0) - const, \quad [M_0] = dyn.cm, \quad (const \sim 10.73),$$

as well as the surface wave magnitude M_S that is based on the magnitude of surface waves, period of the measurement $20\,sec$; the body wave magnitude mb that is based on the amplitude of the direct P wave with period of the measurement $1.0 - 5.0\,sec$ and the Richter magnitude ML assigned a magnitude number to quantify the energy released by an earthquake. There are several other methods for estimation of the seismic moment (Tsuboi et al. (1995), Borman (2002) Aki and Richards (2002)). For application in the tsunami warning Tsuboi et al. (1995) introduced a simple and robust method for determination of M_0 and M_W based on P waves only, that is called as the M_{WP} method. This method can be used with great merit because M_{WP} can be determined within minutes of earthquake detection. There exist also another methods that can be used for tsunami warning on global scales, but can not be used for short distances of the event's origin.

As mentioned above tsunamiearthquakes are generated in the subduction zones, that are characterized by enormous frictional forces between the subducted lithospheric plates and the overriden lithospheric plates. These frictional forces prevent slow and steady rates of subductions and sticks of colliding descending and overriden lithospheric plates. Due to the movements of both colliding plates the deformation energy accumulate in the subduction zone over a long period of time, i.e., over decades or even centuries. Due to forces, frictional and tangential, acting on the contact boundaries of descending and obducting lithospheric plates two situations can occur: (i) firstly, if the tangential forces in the absolute values are less than the frictional forces acting on the contact boundaries, then the frictional forces preclude the mutual motions of both colliding plates and in the subduction zone the deformation energy is further accumulated; (ii) secondly, if the tangential forces in the absolute values reach the frictional forces, then there are no forces that can preclude the mutual motion of both plates and then the contact points change their positions in the directions opposite to that in which the tangential forces act, that produce an earthquake, and moreover, it will be further accompanied with an earthquake or a swarm of earthquakes in the zone of faults. Moreover, the fractured part of the overriden lithosphere uplifted or dropped that this gives also enormous shoves to the overlying water. At the same time, inland areas of the overriding lithosphere are suddenly sunken. These sudden motions and transportation of great part of lithospheric masses cause the tsunami. These ruptures of the overriden lithosphere are characterized by slow ruptures. These slow ruptures are not the physical reason for the efficient generation of tsunami waves. They give indication of the earthquake location, and moreover, they indicate that the tsunamigenic potential in this region may be greater than expected bases on the amplitude of seismic waves at high frequencies. Nevertheless, the estimation of ratio of total deformation energy to high frequency energy or to seismic moment provide a robust measure of the tsunamigenic potential.

Subduction zones are thousands of kilometers long with rupture width of the order of a hundred kilometers, and the ruptures can jump from one section of a subduction zone to another (see the case of the Sumatra earthquake at December 2004). Major earthquakes, the tsunamigenic earthquakes, are large because of large contact surfaces between both descending and overriden lithospheric plates and their mutual slips in the subduction zone, further large rupture length and width as well as rupture slip. Movements of the sea floor as well as movements on the faults are fundamental for tsunami generation. And moreover, these sudden motions are the causes of the tsunami, because they give enormous shoves to the overlying water. At the same moment the sediments with respect to the tsunamiearthquake and the aftershocks can generate submarine mudslides and landslides, that can be also possible origins of a new tsunami waves, and thus as a result they can modified the primary tsunami waves. Moreover, at the same time the volcanic activity as well as typhoon can also generate very high sea waves that can modified the primary tsunami wave amplitudes that will strongly increase.

Seismic monitoring and earthquake warning systems, stress-strain monitoring, volcanic and tsunami warning systems as well as typhoon's warning system will be fundamental for the security of global parts of the Earth's surface in the subduction zones.

1.2.2. Generation and location of tsunamis

Tsunami are series of huge waves that are generated by large-scale underwater disturbances - mass transport events, evoked by the different mechanisms such as (see e.g. Bryant (2001), (2008), Mader (2004), Mosher et al. (2010), Satake (2012), Yamada et al. (2012))

(i) seabed displacements being a result of seismic activity in subduction zones;

(ii) submarine (underwater) landslides at convergent margins and submarine mudslides and debris avalanches;

(iii) underwater explosions;

(iv) volcanic eruptions and volcanic flank collapse at convergent margins and oceanic islands;

(v) meteorite impacts with oceans.

The intensity and frequency of the tsunamis depend upon the sources. Destructive tsunamis are mainly generated by tsunamiearthquakes. The magnitude of tsunamis, that were generated by the tsunamiearthquakes, can be grown with respect to effects of typhoons, submarine landslides as well as of oceanic bed motion activity in these regions. The definition of "tsunamiearthquake" was introduced by Kanamori as an earthquake that produces a large size tsunami relative to the value of its surface wave magnitude M_S.

Maximal number of tsunamis in the world's oceans and seas is situated in the Pacific Ocean and the closed regions and is of about 53 %, in the Indian Ocean (where there are two major subduction zones that can generate tsunamis) is $> 22\,\%$, in the Caribbean region is 13.8 %, in the Mediterranean region is $> 10\,\%$ and the rest corresponds to the Atlantic Ocean regions. In the Pacific Ocean region 82.3 % of tsunamis were generated by the earthquakes, 4.6 % by the landslides, 4.6 % by the volcanic activity and the rest of events is unknown.

Tsunamis are typically generated by earthquakes that occur along the subduction zones. Tsunami waves, that have periods $100\,s$ to $2000\,s$, are generated by the earthquake's mechanism discussed in the next subsections based on the contact mechanics in connection with the fracture mechanics in useful rheologies. The word "tsunami" represents a Japanese word and has a meaning "harbor wave" ("tsu = harbor, nami = wave") and is used by the international communities for a series waves traveling across the ocean with extremely long wavelength that result from seismic or volcanic activities as well as submarine landslides, etc., in the ocean floor. The tsunami's wavelength is the distance between successive crest (or trough). The wave height is the difference between the wave crest and the wave trough. The wave steepness is defined as the ratio between the wave hight and the wave length. The amount time between successive waves is the wave period. The first tsunami waves in some cases are not the large, but the subsequent waves can be many times larger. The speeds of these type of waves are given by

$$c = \frac{L}{T} = \sqrt{\frac{gL}{2\pi} tanh(\frac{2\pi d}{L})}, \tag{1}$$

where c is the wave speed, L is the wavelength, T is the wave period, d is the water depth, g is the gravity acceleration and $tanh$ denotes the hyperbolic tangent. The lengths of these waves are very long, the wave amplitudes increase enormously in the same way as the wave lengths decrease. If the water depth d is larger than the half of the wavelength, then the waves are referred to as deep water waves, if the water depth d is between $\frac{L}{20} < d < \frac{L}{2}$, then the waves behave as intermediate water waves, and if the water depth d is less than $\frac{L}{20}$, then the waves correspond to shallow water waves. Then we have

$$\begin{aligned} c &= \sqrt{\frac{gL}{2\pi}} \quad \text{for the deep water waves,} \\ c &= \sqrt{gd} \quad \text{for the shallow water waves.} \end{aligned} \tag{2}$$

In the deep ocean, tsunamis can propagate at more than $900 km/h$, and in the shallow water they propagate much more slower than in the deep water, i.e., of about $20 km/h$. Moreover, tsunamis propagate on the large distances with limited energy losses, and therefore, they have sufficient energy to propagate across the oceans. Tsunami waves move outwards away from their origins, that is, they move outwards perpendicularly to the subduction trench, where they were originated. The factors that influence the magnitude of tsunamis are the magnitude of the earthquake, the shape and orientation of the subduction zone as well as the other effects that were evoked by e.g. typhoons, landslides, mudslides, volcanic eruptions, etc. Therefore, the mathematical modelling of tsunami waves is the crucial point for determination of the tsunami risk assessment. For more details about the ocean dynamics see e.g. Kowalik and Murty (1993), Mader (2004).

We saw in the previous subsection that for the estimation of the earthquake size the seismic moment M_0 is acceptable, that combines the fault area S, the amount of slip d and the shear modulus μ, that is, $M_0 = \mu.S.d$, or the moment magnitude $M_W = \frac{2}{3}log_{10}(M_0) - 10.73$, where M_0 is expressed in $dyn.cm$. For example, for the large earthquake in Chile May 22, the seismic moment was determined by $M_0 = 2.5 \times 10^{30} dyn.cm$ and the moment magnitude by $M_W = 9.55$. New methods based on the GPS measurements seem to be very useful for rapid determination of the earthquake size and the tsunami potential (Blewitt et al. (2006)). The tsunami magnitude scale M_t was introduce by Abe (1979),(1981), and it allows estimation of earthquake magnitude based on tsunami wave amplitudes measured on the gauge instruments, that is,

$$M_t = log H2 + log X + 5.55,$$

where $H2$ is the maximum crest-to-trough amplitude on tide gauge records in meters and X is the distance from the epicenter to the recording station along the shortest oceanic path (in km). It can be mentioned that M_S (i.e., surface wave magnitude) can be smaller than the tsunami magnitude M_t. Bryant (2008) published the relation

$$m_{II} = log_2 . H_{max},$$

where m_{II} is the tsunami magnitude and H_{max} is the maximal hight of the sea level.

Probable occurrence of the tsunami origin on the value of earthquake magnitudes (in %) for the magnitudes > 7.3 is 100%, for the magnitudes in the range $7.0 - 7.2$ is 67%, for the magnitudes in the range $6.7 - 6.9$ is 17%, in the range $5.8 - 6.2$ is 1.4% and for the magnitudes < 6.2 is less 1.4%. There exist six degrees of tsunami sizes, denoted I to VI, from very small tsunamis that are recorded on the tsunami meters up to VI, characterized by the catastrophic tsunamis. In Table 1 the values between the earthquake magnitude M (Richter scale), the tsunami magnitude m_{II} and the maximal sea level L in m are presented (modified after Bryant (2008)).

Table 1. Values between the earthquake magnitude M (Richter scale), the tsunami magnitude m_{II} and the maximal sea level L in m

earthquake magnitude M	*tsunami magnitude*	*max. sea level in* m
6.0	-2	<0.3
6.1-6.5	-1	0.5-0.7
6.6-7.0	0	0.75-1.5
7.1-7.5	1	1.6-3.0
7.6-8.0	2	3.1-7.0
8.1-8.25	3	8.0-12.0
8.3-8.5	4	16.0-24.0
8.6 *and more*	5	>32.0

Large tsunamis are generated by large earthquakes with $M > 7$, which release seismic energy at very long periods, therefore, the correct size of these earthquakes from the point of view of seismic moment can be determined from the measured very long periods. Seismic and sea level data are needed to detect a tsunami occurrence. These data are requested by the National Tsunami Warning Centers ($NTWC$) and the Regional Tsunami Watch Providers ($RTWP$). Knowing the type and the strength of an earthquake is fundamental for determination of its potential to generate a tsunami. Therefore, seismic data play a main role in the tsunami warning system. To determine the accurate moment magnitudes, both institutes - $NTWC$ and $RTWP$, require reliable frequencies, low noise, high dynamic range, digital seismic data in real time. These data are also needed for the use in real time in the international tsunami warning systems with minimal delays. Moreover, the tsunameter buoys to detect a tsunami wave signal are used. The National Oceanic and Atmospheric Administration (*NOAA*) introduced the "Deep-ocean Assessment and Reporting of Tsunami ($DATR^{TM}$) buoys situated in the Pacific Ocean and the Atlantic Ocean (and now also in the Indian Ocean are prepared) and at present the NOAA developed its new version $DART-II$, the second-generation of the *DART* system, based on an anchored seafloor bottom-pressure recorder (*BPR*) and a series of moored surface tsunameter buoys for real-time communications with centers, where the *BPR* detects water pressure with a resolution of approximately $1\,mm$ of sea water, with $15-$second average samples. The data then are transported through an acoustic link to the surface buoys. The *DART Project* is a part of the U.S. National Tsunami Hazard Mitigation Program for the early detection of tsunamis and for real-time information in the whole open ocean. For coupling tsunami warning the coastal sea level tide gage data are used. The *DART* system detect all the type of tsunamis generated by tsunamiearthquates, by volcanic activities, landslides or mudslides as well as of waves generated by typhoons. The satellite monitoring and evaluations of various natural phenomena as the Earth's surface; satellite observations of different aspects of typhoons; satellite data include surface winds, sea surface heights and temperatures, rainfall intensities, atmospheric pressure, locations and sizes of storms and typhoons, clouds covers and their temperatures and further developments, etc., are made by the National Aeronautics and Space Administration (*NASA*), that also elaborated all obtained data. For these observations *NASA* has several satellites in orbits around the Earth that are used to study different aspects of tropical cyclones.

At the end we introduce several world largest earthquakes with ensuing tsunamis. Firstly, there is the Chile earthquake and tsunami of 22 May 1960 in the coast of South Central Chile with Moment magnitude $M_W = 9.5$ with the foreshock situated in the continental lithosphere, with the main shock in the focal depth $33\,km$ of depth and its magnitude was estimated of about $2.7\,10^{30}\,dyn.cm$ (with aftershocks that continued for weeks) (Kanamori and Cipar (1974)), that generated one of the most destructive Pacific-wide tsunami with tsunami wave hight of about $25\,m$. The tsunami was destructive in Chile, Hawai and Japan and the closed regions. This earthquake was accompanied with a series of earthquakes that began May 21 1960 and continued by a series of earthquakes in southern Chile extending over a $1,300\,km$ long zone, between 37 and 48 degrees south latitude. The second one was the Great Alaska Earthquake March 1964, the second-largest earthquake in the modern history, with the magnitude $M_W = 9.2$, situated in the south-central Alaska and situated $800\,mile$ away from Anchore, and Seattle, Washington, more than $1,200\,miles$ to southeast of the fault rupture. For information the Japan's 2011 Tohoku Earthquakes with magnitude $M_W = 9.0$ is the fourth largest in the world since 1990. The earthquake was situated in northern Honshu, Japan.

1.2.3. Landslides and mudslides, their locations and mechanisms of their origins

Giant submarine landslides and mudslides result from falling of unconsolidated and eroded submarine sediments with respect to the earthquakes. Submarine landslides are characterized by unstable slopes of rapid sedimentation, of bottom current erosion of the sedimental materials at the base of the delta slope, additional loading at the top of the delta slope, and above all of the earthquake shocks. These earthquake shocks cause loss of the soil strength and increasing in the downslope force on the soil mass. Submarine landslides and mudslides may produce waves large enough to evoke tsunamis, and moreover, to evoke ensuing damages, and destruction of large landscapes of sea coasts. The submarine landslides and mudslides are located in places of deltas of major rivers or streams, and moreover, in places of the sea coasts created by soil rocks, clearing of the land, etc. The stabilities of the submarine slopes depend on the steepness of the slopes and the slope aspects; on composition of the soils and the rocks; on anthropomorphous properties of rocks.

In order to analyze and evaluate the potential hazard related to unstable submarine rock slopes related to consequences of erosion and earthquakes and shocks, it is essential to understand the processes and mechanisms driving the slope instability. Creep movements play an important role in the landslide origin (Keating et a. (2000), Nedoma (2010)), because the creep is a viscous effect.

1.2.4. Processes in the atmosphere, the locations and the origin of typhoons

The instability of the atmosphere with the zone of low pressure and surrounded by air masses with the normal or elevated pressure - known as anticyclone, form tropical cyclogenesis, the tropical cyclones. Tropical cyclones with maximum sustained winds with speeds less than $61.2\,km/h$ ($17\,m/s$ or $34\,kt$ or $39\,mph$) is denoted as tropical depressions. By the tropical storms are denoted systems of strong thunderstorms with surface circulations and sustained winds between $61.2\,km/h$ and $119\,km/h$ ($33\,m/s$ or $64\,kt$ or $74\,mph$), that represent a Category 1 of cyclone. During the end of this wind-speed interval the cyclonic shape start to originate, but an eye of cyclone is not yet originated. In the Beaufort scale the wind speeds are stated as $54\,km/h$ and $102\,km/h$ and in the Staffin-Simpson scale as $54 km/h$ and $117\,km/h$. For winds with speeds more than $118\,km/h$, then we speak about

- the typhoon, in the northwest Pacific Ocean west of dateline;
- the hurricane, in the North Atlantic ocean, the Northeast Pacific Ocean east of the dateline, or the South Pacific Ocean east of $160°E$;
- the Category 3 of cyclone or the severe tropical cyclone, in the Southwest Pacific Ocean west of $160°E$ or Southeast Indian Ocean east of $90°E$;
- the tropical cyclone, in the Southwest Indian Ocean;
- the very severe cyclonic storm, in the North Indian Ocean.

The word typhoon was derived from the Cantanese word "tai-fung" denoting a great wind. The typhoons are huge whirlwinds with air circulation around an area of low pressure and extending about $100\,km$ from the center, and are formed between $5-30$ degrees latitude, and $110-180$ degree longitude. Typhoons develop over ocean surfaces having temperatures at least $26.5°C$, throughout a sufficient depth, that is needed as a fuel for the cyclone process, and a rapidly cooling atmosphere with height such that it is potentially unstable to moist convection. Therefore, the fundamental conditions for the typhoon origins are warm oceanic waters, moisture, and a wind pattern near the surface of the ocean that spirals air inward. Thus, the eye - the center of a typhoon, is relatively calm. The great activity is in the area around the eye, the eye-wall with $16\,km$ to $80\,km$ of wide.

Maximum sustained winds are estimated of about $220\,km/h$ to $315\,km/h$. At the top of the eye-wall, the air is propelled outward, increasing the air's upward motion. One part of the air moves inward and sinks into the eye. The typhoons, hurricanes and tropical cyclones use the energy from the ocean in the form of sensible and latent heat (enthalpy) fluxes (Price (2009)). The cooling of the ocean surface reduces the typhoon-ocean heat flux and thus reduces the typhoon intensity. If the temperature of the ocean surface water fall bellow $26.5°C$, typhoons as well as hurricanes and tropical cyclones dissipate, and as a result, typhoons as well as the other tropical types of cyclones help to maintain equilibrium in the Earth's troposphere as well as to maintain a relatively stable and warm temperature worldwide. It is evident that in the evolution of all types of tropical cyclones the Coriolis' force plays an important role.

In the Pacific Ocean besides the typhoons also the super-typhoons that have gained wind-speeds of above $150\,mph$ and more, are observed. This types of typhoons are equivalent to a $4th$ category of hurricanes. For example, the Super Typhoon Haiyan in the central Philippines packed sustained winds of up $305\,km/h$ ($190\,mph$). Their surge heights are $4-10\,m$ and more and pressures are less than $920\,hPa$.

1.3. Proposal of the Research Project

Regions with very active seismicity, represented with frequent seismic events and bigger earthquakes, with ensuing tsunamis as well as with bigger climatic events such as typhoons and ensuing deluges and landslides represent one of the most detrimental natural hazards. These events have been responsible for some of the most destructive natural disasters in terms of human and economical losses. Therefore, the aim of the main part of the global research project for the Japan region, and the other closed Pacific regions, is to give an idea how to predict the future situations in such endangered regions and how to predict such future situations before the future events, and therefore, to prevent their consequences. Tsunamis as well as typhoons reshape the geology near the coast of the sea by eroding sand from the large shore areas. The large parts of landscapes are reached by enormous quantities of water and ensuing deluges and landslides of soils and with rocks saturated by a water. Moreover, these regions are handicapped by the effects of earthquakes and their immediate consequences. Therefore, the prediction of some reached landscapes with potential landslides and mudflows are necessary. Such possibilities allow stress-strain measurements in situ, that is, in deep boreholes, based on the overcoring method (for determination contemporary stresses), and long-term stress-strain monitoring in the same places, based on the string tensometry or on the more accurate optical fibre tensometry. From this point of view the research project will have several main research subprojects:

(i) monitoring of seismicity in the region of interest with the earthquake early warning and the stress-strain monitoring in situ;

(ii) monitoring of the tsunamis and determining the type of their origins as well as determination the amplitude (height) and the wavelength of the propagated tsunami wave;

iii) monitoring of climatic and thermo-hydro-meteorological data for the typhoon forecasts and for determination of the waves evoked by the typhoon and their propagation in the direction to the continent and the islands and across the islands and the continent;

(iv) monitoring of the ambient stability of natural land forms and soils and rocks in the landscape;

(v) monitoring volcanic activity in the regions of interest;

(vi) monitoring all of remaining geophysical fields such as magnetotelluric field, hydrology and geology of the region of interest and their changes in connection with watering of rock materials in the region, etc.

(vii) mathematical simulations of all processes taking place in the region of interest, based on the useful rheologies.

The idea of the proposal of research sub-projects, utilized all observed data from the points (i)-(vi), is based on the mathematical simulation of plate tectonics in the region of interest with ensuing propagation of seismic waves, on mathematical simulation of coupled thermo-mechanical and thermo-hydro-dynamic processes in the Japan region as well as in all other regions in the Pacific ocean area.

(A) Stress-strain observations in situ and mathematical simulations of tectonic processes in the endangered regions

This proposal deals with a stress-strain monitoring in situ and the mathematical plate tectonic models and the models of seismic wave propagations based on the multibody contact theory. The obtained results can be used for the studies concerning with the Japanese (and the other) subduction zone(s) and the earthquakes in this region and their ensuing consequences.

Knowledge of the state of stresses inside the Earth is fundamental for understanding of all tectonic processes in regions of interest. It is also of importance to the design of large structures like the nuclear power plants or the radioactive waste repositories, reservoir dams, etc., namely in regions endangered by great earthquakes with ensuing tsunamis and (super-)typhoons.

Stress-strain measurements:

Rock parameters which are to be measured or monitored by field instrumentations in connection with the problems related to the project of in situ stress-strain determination can be broadly classified as:

(i) rock mechanics site constants like in-situ stress, in-situ strain and in-situ strength of rock mass,

(ii) variable geodynamical and geomechanical parameters like ground movements and variation in stresses and strains,

(iii) parameters to evaluate performance of different strata control systems like pressure monitoring, etc.

To determine the present values of stress-strain fields at different locations in boreholes, the hydrofracturing technique and the overcoring method can be used (Nedoma (1998)). For the overcoring method the 3D string tensometry or the 3D tensometry based on optical fibres can be used with a great advantage.

Mathematical models:
The models must be formulated as the *3D* models. The mathematical models will be based on the linear (thermo-)visco-elastic rheology with short or long memories and/or on the nonlinear (thermo-)elasticity or on the nonlinear (thermo-)visco-elasticity with short memory or on the (thermo-)visco-plasticity and on the theory of multibody contact problems. Such model problems allow to simulate tectonic processes taking place in the upper parts of the Earth (i.e., in the lithosphere and the upper mantle) and based on the plate tectonic hypothesis and the theory of multibody contact problems in the above mentioned rheologies, as well as allow to simulate ensuing earthquake origins and ensuing propagations of seismic waves in the regional lithosphere broken up into several greater and smaller geological blocks.

For the used models the following professional demands must be requested:

- the geometry of the investigated region determined from the digitized cartographic maps of the investigated regions,
- the geometry of contact surfaces between geological blocks and lithospheric plates, determined from the geological and geophysical observations and analyzes,
- material properties of rocks of the investigated regions, i.e., the input physical data as elastic and viscous constants, friction coefficients, etc.,
- acting loading forces,
- temperatures and other needful thermal data.

Then the obtained in-situ data will be compared with results obtained from the mathematical modelling that will be discussed later, and moreover, the data from the deep boreholes by mathematical models extended into the whole region of interest, that can be endangered by the earthquake, aftershocks as well as ensuing tsunami and by the typhoon's waves. From this point of view it is evident that this project of stress-strain monitoring in situ and mathematical modelling of stress-strain fields must be closely connected with the project on the "Security of Countries Endangered by Huge Tsunami and Typhoon" and with the project concerning with "Predict the Possible Endangered Regions During the Huge Tsunami and the Typhoon."

(B) Proposal how to predict the future situations in the endangered regions after huge tsunamis and/or typhoons

The idea of the proposal how to predict the future situation in the endangered regions is based on the mathematical simulation of coupled thermo-hydro-mechanical and thermo-hydro-dynamical processes together with the oceanic and climatic processes in the endangered region during huge tsunamis, typhoons and consequences of enormous quantities of water in these regions onto a stability of slopes in these regions and onto the security of these regions. The great Japanese regions and the closed Pacific regions under the tsunamis and typhoons and areas with non-stable seashores, seasides, wharves, lakes and dams represent the characteristic reason for application of such model situations in practice.

The rheology used in our models will describe with high accuracy situations in endangered regions with respect to all effects of tsunamis (evoked by tsunamiearthquakes) and great typhoons, to bigger storms with great fall-out of water, resulting deluges and the ensuing relevant landslides.

The main goal of the presented idea is to connect the oceanic observations, climatic observations and the corresponding climatic models with the thermo-hydrodynamic and the thermo-mechanic models, with the possibility to estimate future destruction of these endangered regions with landslides of the non-stable slopes.

The integral part of this idea is to determine the causes of risks in endangered regions and the determination of characteristic parameters for their evaluations and analyzes.

The main goal is to help with specification and categorization of all these cases with aims at removal relevant risks and symptoms of a crisis in such critical situations. Such complex evaluations of the future situation then start to be an important foundation for elaboration of the "Analyses of risks" for critical situations in such endangered regions, because that can be simulated before future events, and therefore, such analyses of risks can reduce the future destructions of the shore (coastal) landscopes.

Requirement for the input information and data:
The main goal is to simulate and to predict all effects of the thermo-mechanical and thermo-hydro-dynamical processes in the unsafe regions, intervened by tsunamis evoked by great earthquakes, typhoons, etc., in critical situations.

Mathematical models:

The mathematical models discussed will be based on the non-linear (thermo-) visco-elastic rheology and on the (thermo-)visco-plastic rheology of Bingham type coupled with tsunami's, climatic and hydrological models. The models must be formulated as the $3D$ models.

The non-linear (thermo-)visco-elastic rheology is based on the special type of the strain energy density function, that will be a function of the small strain tensor of the form $E = A^{\lambda}(e_{ij})$, where A is a scalar-valued function of strains e_{ij} and λ is a positive parameter. By $\lambda = 1$ we model the classical (thermo-) visco-elastic rock materials, by $\lambda < 1$ we model the softened rock materials and by $\lambda > 1$ we model the hardened rock materials. Therefore, the parameter λ simulates the degree of non-linear properties of rock materials.

The Bingham rheology is characterized by two parameters - the threshold of plasticity $\hat{g}$ and threshold of viscosity $\hat{\mu}$. The threshold of viscosity $\hat{\mu}$ has the same meaning as in the case of the Newtonian liquid, the threshold of plasticity $\hat{g}$ is from the interval 0 to ∞. For $\hat{g} = 0$ the environment is described by the liquid, for very small $\hat{g}$ the environment is characterized by the liquid saturated more or less by minerals and rocks and for $\hat{g} \longrightarrow \infty$ we have absolutely rigid materials. Therefore, between $\hat{g} = 0$ and $\hat{g} \longrightarrow \infty$ we can model all types of rocks, which are loaded by external forces as loading forces induced by the typhoon, weights of rainfalls, weights of materials in the rockcovers, weights of industrial buildings, etc. The values of $\hat{g}$ must be determined from the mechanical properties of rocks in the investigated regions. In general the threshold of plasticity (yield limit) $\hat{g}$ is a function of the location, mass density ρ and time as well as of a water concentration. The Bingham solid investigates to describe the deformation, the displacement velocity and stress-strain rate fields. The inhomogeneous yield limit $\hat{g}$ represents the fundamental property for a stability of a natural submarine slope. It is evident that the property of blocking ensures with a great advantage the stability of submarine slopes, because the solid is blocked in its natural configuration and the beginning of a flow in a rock massif represents the beginning of the submarine landslide or of the submarine mudslide, when the rocks were saturated by a very much of water.

From the characteristic properties of the used rheologies the following professional demands are requested:

- the geometry of the investigated region will be determined from the digitized cartographic maps of the investigated region,
- the material properties of the rock massif in the investigated region, i.e., the input physical data as elastic and viscous coefficients $c^n_{ijkl}, n = e, v$, thresholds of plasticity and viscosity $\hat{g}$, $\hat{\mu}$, temperatures, etc. will be determined from the geological, geophysical and hydro-dynamical observations and from the laboratory experiments,
- the acting loading forces will be determined as resulting data from the climatic, geological, geophysical and hydrological observations and studies as well as loading forces developed as the weight of sediments, as the weight of the industrial objects, etc.,
- the temperature and the hydro-dynamical data, like flows of water in the rivers in the investigated region, typhoon data and data of its evolution in a future, an absorbing capacity of water in rocks, properties of rocks in connection with stability of slopes, etc., will be determined from hydro-geological observations and studies,
- influences of the sea surges of the rushing up typhoon's and tsunami's waves in the presented model will be modeled by the boundary value conditions and they will be determined from climatic observations and studies. In more detailed studies these models will be based on the coupled model problems simulated typhoon's and tsunami's waves.

The result of the presented models will be in distributions of the stress-strain fields, velocity and displacement fields in the investigated region, their horizontal and vertical components, thermal field, estimations of the passage of flood waves, estimates of quantity of the running water in the investigated region, estimates of the quantity of water that can be intercepted by dams and swamps, estimates of possible waterlogging, and therefore, the estimation of possible unstable slopes and prediction of their incidental landslides during the critical situations, etc.

It is evident that cooperation of specialists from oceanology, climatology, geology, hydrology, volcanology and mathematics, namely of specialists from the workplaces of the Japan Meteorological Agency (JMA); Meteorological Research Institute, an associate institute of the JMA, which is engaged in analyzing and predicting meteorological, geophysical, hydrological and oceanological phenomena as well as extensive related technology; the National Typhoon Center of the JMA; the National Meteorological and Hydrological Services, specialists from the workplace of the Japan Tsunami and Earthquakes Centers as well as co-operation with the Joint Typhoon Warning Center (JTWC), the National Hurricane's Center in Miami, and the other centers in the Japan and in the Pacific region, are fundamental for successfully close of the project. The similar cooperations are also necessary in the regions of the Atlantic Ocean as well as in the region of the Indian Ocean.

The main goal of this chapter will be to give mathematical models and then mathematical and computational methods and algorithms that will be useful for description and then for solutions of corresponding mathematical model problems. These model problems will describe the origins of earthquakes, ensuing submarine landslides, mudslides and rock-falls, the effects of the submarine volcanic eruptions and then the evoked ensuing tsunamis as well as the effects of the typhoon. And in the consequence of all these types of events, then they will model the resulting consequences of propagated tsunami's and typhoon's waves on the adjacent continental landscapes.

2. Mathematical Models, Computational Methods and Algorithms

2.1. Introduction

Most of our knowledge about subduction zones comes from knowledge of the earthquake hypocenters and the focal mechanisms (Isacks and Molnar (1971), Uyeda (1982), Aki and Richards (1980), (2002) and many others). The mechanics of subducted lithosphere of great lithospheric plates and blocks are connected with the geodynamic processes inside the Earth.

In this section we will present a brief summary about the origins of earthquakes and a brief summary how the tsunamis are generated by the earthquakes. Because at the same time the submarine landslides, mudslides, rock-falls, volcanic eruption as well as the typhoon can be occured in the distressed oceanic region we will introduce some information that can be useful for the mathematical models and that will be useful to describe all of these mentioned events. Then we will present some of mathematical methods (based on the suitable rheology) that can describe the origins of the earthquakes and the ensuing other above mentioned events, and moreover, the generation and evolution of ensuing tsunami in the region under the typhoon. Next, we will shortly describe the main mechanisms of the earthquakes and the ensuing tsunamis, landslides (and so on) under typhoons and their consequences for large landscapes in the distressed region.

Earthquakes are assumed to originate

(a) from a single point - the focus or the hypocenter, that is situated in the crust or in the lithosphere within about 80 km from the surface for the shallow and middle earthquakes, and up to depths about 700 km from the surface for the deeper and very deep earthquakes;

(b) by the movement along a fault plane originated during fractures in the lithosphere, so that the focal area may extend for several kilometers, and

(c) by movement along contact surfaces of subducting and obducting lithospheric plates and blocks in the subduction zone, so that the focal area may extend for several ten to several hundred kilometers of length, which can be accopanied by fractures of several kilometers of length.

The point on the Earth's surface vertically above the focus is denoted as the epicenter and the angle subtended at the Earth's center by the epicenter and the point at which the seismic waves are observated (the points where the seismographs are situated) is denoted as the epicentral angle. The strain energy released by an earthquake is transmitted throughout the Earth by several types of seismic waves (e.g. P, S, Love, Rayleigh, etc., waves). The compressional or longitudinal or P waves propagate by the velocity

$$v_P = \sqrt{\frac{k + 4/3\mu}{\rho}}, \tag{3}$$

where ρ is the density of rocks, k is the bulk modulus of the rocks, μ is the shear modulus or the rigidity of the rocks. The shear or transverse or S waves propagate by the velocity

$$v_S = \sqrt{\frac{\mu}{\rho}}. \tag{4}$$

While the P waves correspond to elastic deformation of rocks by compression/dilatation (the particles of the transmitted rock oscillate in the direction of the wave travel) so that the disturbance proceeds as a series of compressions and rarefactions, the S waves correspond to elastic deformation of rocks by shearing (the particles of the rock oscillate at right angles to the direction of propagation, therefore, we distinguish two types of shear waves, the S_H waves (polarized in the horizontal plane) and the SV waves (polarized in the vertical plane). The P waves propagate also in the liquids, while the S waves cannot propagate in the liquids because the rigidity μ of a liquid is zero.

The earthquakes are detected by seismographs, that are the instruments that detect small ground displacements, velocities, or accelerations associated with passage of seismic waves. The seismographs create the local and global network of seismograph stations for monitoring seismic activities. In the world there exist the World-Wide Standarized Seismograph Network (WWSSN) that is based on the analogue instruments and the Global (Digital) Seismograph Network (GSN), based on the digital instruments. By the identification of the seismic arrivals (phases) on the seismograph records the earthquake's location can be determined.

There are several methods for determination of the earthquake's location. Earthquakes that occur at large (teleseismic) distances from a seismograph are located by the identification of seismic arrivals (phases) on the seismograph records. Due to the different velocities of seismic waves (P and S) their arrivals are also different in time. The focal depths of teleseismic events are determined from the measured times of arrivals of the dirrect phases P and pP, where the pP phase corresponds to a short path multiple event which follows a similar path to P after first undergoing a reflection at a surface of the Earth above the focus. Hence $P - pP$ time difference is a measure of focal depth. Another method used to determine focal depth is based on the sP phase, because S wave is reflected as a P wave from the Earth's surface at a point near the epicenter. The depth of an earthquake can be determined from the sP phase in the same manner as in the previous pP phase, using the appropriate travel-time curves, or the depth tables for sP. These methods are used for foci at depths of less than $100\,km$ as the $P - pP$ time separation becomes very small. The focal depth of local earthquakes can be determined if a network of seismographs exists in the vicinity of the epicenter. Then, for each station recording the earthquake, a model for the velocity structure of the Earth is used, that will be applied for computation the travel path of the seismic wave from the focus to the station, and therefore, to compute the direction in which the seismic wave left the focal area. These directions are then plotted and further elaborated for detailed analyzes. For details see some special seismological books (e.g. Båth (1979)).

The seismic sources are of three types based (i) on the fractures occurred in the lithospheric plates and the Earth's crust in the sense of the fracture mechanics, (ii) on the multibody contact mechanics, formulated as fracture (fault) problems with non-penetration conditions and with or without Coulomb friction, and (iii) on the multibody contact mechanics with Coulomb friction, that describes movements along the contact boundaries of the subducting and the obducting lithospheric plates in the subduction zone in the sense of the plate tectonic hypothesis. These two last types of seismic sources will be studied later. The results obtained in the dynamic contact mechanics and the elastodynamic fracture mechanics can be also applied to the analysis of geodynamic processes.

The analyzes of the crustal earthquakes and deep earthquakes in the lithosphere are caused by sudden movement on faults, which were arisen by broken upper lithospheric plates into a great number of geological blocks. Fracture mechanics is concerned with the analysis of the stability of cracks. With regards to the analysis of the stresses or displacements associated with crack growth or relative displacement on the surface of a discontinuity the two approaches are coincident. The analyzes of fracture problems, based on the fracture mechanics, have studied with analyzes of earthquake's mechanisms and earthquake prediction (Rice (1980), Aki and Richards (1980), Gdoutos (2005)) as well as plate tectonic approaches (Segall (1984), Nedoma (1998), Cox et al. (2005)).

Deformation processes occur in the Earth over a wide range of time scales. At one extreme there are very slow deformations caused by the large-scale tectonic motions of the lithospheric plates with ensuing micro-cracking, inter-crystalline fractures and macro-cracking, and on the other hand there are dynamic motions and fractures associated with earthquakes. All fractures in the macro-scopic and the micro-scopic scales are dynamic. The dynamic nature of fractures in the macroscopic world is manifested when the inertia of relatively large pieces of rock materials is large enough that the correct balancing of the energy of fracture requires including kinetic energy. For a fracture that is already propagating, we may say, that the dynamic (inertial) effects are important when the fracture tip speed is small in comparison with the stress wave velocities. But even when these dynamic effect conditions at the macroscopic (i.e., geological) scale are not met, still at the microscopic (i.e., atomic) scale, the inertia of individual atoms must be accounted for their separation from one another, even when the fracture at the macroscopic scale appears to be advancing quasi-statically. Bond fracture is a dynamic process. Cox et al. (2005) studies the dynamic fractures from the geological macroscopic and microscopic scales point of views, because the dynamic fracture problem is fundamental in the earthquake analyzes.

The dynamic effusion and ejection of magma that occur during a volcanic eruption deal with the dynamic fracture of high viscous rock materials and with the driving forces for fractures provided by the pressures in diffusely distributed bubbles. The rate of decompression then will affect the character of the damage developments and the natures and the evolutions of volcanic eruptions.

The qualitative models of earthquakes are developed for the dynamic propagation of fractures and resulting propagation of deep faults and ensuing emanation of an accumulated deformation energy in the form of the earthquakes. The models will be formulated more generally, facilitating formulation of problems concerning crack propagations as well as propagation of resulting seismic waves propagated in rocks occupying the region of our interest. This region will be determined by the parameter functions $\alpha(\mathbf{x}) \geq 0$, which physically have a meaning of damping. We will introduce an inhomogeneous absorption (damping) parameter function α, such that it corresponds to the physical reality in the studied region, and moreover, such that in the case of seismic wave propagation it becomes completely absorb propagated waves at a certain distance D near the boundary (we denoted it as ${}^1\Gamma_u$) with Dirichlet's displacement boundary conditions (zero boundary conditions), allowing to study propagation of seismic waves in a certain finite domain of interest (Nedoma (1998)). Since the waves are completely absorbed in the region near the boundary ${}^1\Gamma_u$, then none of seismic waves are reflected back. The presented method allow us to simulate propagation of fractures, which extend to faults and then originate earthquakes with ensuing emanation of accumulated deformation energy into the neighboring region in the form of seismic waves. The same idea using the damping function $\alpha(\mathbf{x})$ can be applied in the case if the earthquake is evoked by the mutual movements of the lithospheric plates or the lithospheric blocks in the subduction zones in the sense of the plate tectonic hypothesis. All these discussed situations will be simulated by the suitable algorithms and will be discussed later.

The rocks of the upper lithosphere under low temperature and low pressure are in the brittle state. A rock material is said to be a **brittle material,** if, when subjected to the stresses, it breaks without significant deformation and it absorbs little energy prior to fracture, even those of high strength. The rocks under higher temperature and pressure change their mechanical properties and start to be (visco-)plastic. From the fracture mechanics and the mathematical point of views principal difficulties of fracture problems are connected with the non-regular character of boundaries caused by the presence of a fracture within a domain and with the stability of fractures. The formulation using non-penetration conditions was introduced for approximation the real situation at the fracture faces. Such formulated fracture problems lead to solve the variational inequalities. The real situations at the deep fault faces are much more complicated due to fragmentation and melting processes.

Elastodynamic analyzes (Atkinson (1987), Gdoutos (2005)) of fracture problems in rocks indicate that stresses and displacements evoked by dynamic loadings are greatly differ from those associated with the corresponding static loadings. Therefore, the dynamic stresses are higher than the corresponding static stresses. This reality can be explained by the intersection of elastic waves with the crack faces and other characteristic boundaries of the rock body. Moreover, the mechanical properties of rocks depend also on the time for which the applied loadings are maintained in the rocks. Dynamical loads evoke high stresses near cracks and fractures take place so rapidly that there is insufficient time for yielding to develop. Energies are therefore released within a short time that lead to rapid crack propagations, that explain brittle fractures in rocks. The dynamic fracture mechanics problem are of two types: (a) the first one arises when a rock with a stationary crack is subjected to rapidly varying loads e.g. impact and impulsive loads; these problems are related to initiations of rapid crack growths, crack speeds, crack branching and/or crack arrest; (b) the second one are related to the situations where cracks reach a corresponding point of instability and move rapidly under slowly varying applied loading. Therefore, into these types of dynamic problems belong:

(i) dynamic problems evoked when rocks with stationary cracks are subjected to rapidly varying loads, that is, impact and vibration problems, problems in which at inhomogeneities the dynamic stresses are higher than the stresses observed in corresponding static cases, and such that for a crack the intensity of stress field in the vicinity of the crack tip can be affected by dynamic effects, that is, there are cases for which fractures at crack tips do not occur under an applied system of loads, but where cracks propagate when the loads are rapidly applied and give rise to waves, which strike the crack tip; and

(ii) earthquake mechanism problems, problems in which rapid crack growths are possible, and in which dynamic effects affect the stress field near rapidly propagating cracks, and thus give the conditions for further unstable crack propagations or for their arrests.

The useful method for the crack propagation analysis is the quarter-point method (Whiteman and Akin (1984)). The method makes use of the "r^{β_i}, $\beta_i > 0$", singular form and produces an approximation which exhibits this form. The method is discussed in detail in Whiteman and Akin (1984). The $r^{\frac{1}{2}}$ form can be obtained with the quadratic 2D elements by moving mid-side nodes on the element sides passing through the singular point to points on these sides, one quarter of the side length from the singularity; hence the quarter-point and thus the quarter-point method. The quarter-point elements have been extended to three dimensions to deal with line singularities. The quarter-point adaptation of the fifteen-node prism element is one of such elements. The twenty-node isoparametric brick element has been similarly adopted. This method is discussed in details in Thompson and Whiteman (1985). The finite element methods as the $h, p, h-p, m, h-m, h-p-m,$ etc., methods can be also used in the fault analyzes in the crust and the lithosphere (Nedoma (1998), Šolín et al. (2004)).

From the observation of loading rocks we see that a propagating crack tip acts as an energy sink, and that the mechanical energy is dissipated during the fracture process even through the rock are perfectly elastic rocks. In general stress intensity factors for Mode II and Mode III can be similarly defined. It is evident that, in general, the stress intensity factor k depends also on the geometry of the cracked rock, the applied loading, and the crack speed in the case of a moving (propagating) crack. Formation of a new fracture surface is accompanied by the loss of mechanical energy. This becomes possible if the work of the cohesive (internal) forces is considered that are removed as new fracture surface, which is free of tractions. The cohesive tractions, that are removed, are opposite in direction to the relative displacements of the newly formed fracture surfaces. The resulting work is negative. Thus, mechanical energy is extracted from the rock body. The region over which cohesive tractions are released as the crack propagates is infinitesimal, namely, the tip of the propagating crack. The energy release rate G has a finite value. The relation between the energy release rate G and the stress intensity factor k for a moving crack depends on the crack speed (Atkinson (1979), Gdoutos (2005)).

Mathematical modeling of elastic wave propagations play an important role in seismological studies (Dahlen and Tromp (1998), Seriani (1998), Komatitsch and Tromp (1999), (2002a,b), (2003), Komatitsch et al. (1999), (2000a,b,c), (2003), Capdeville et al. (2003), Chaljub et al. (2003), Liu (2006) and many others). Mathematical model of elastic wave propagations through the partly fractured media is studied in Nedoma (2013). In this study the mathematical model of the Earth with the broken lithosphere under self-gravitation is formulated and the numerical method and an idea of algorithm for a computational realization are presented.

In this study we will assume the existence of contact boundaries between the neighboring lithospheric plates and the neighboring lithospheric blocks in the broken up lithosphere as well as the existence of the contact boundary between the lithosphere and the mantle. Because from the plate tectonic point of view the lithospheric plates and lithospheric blocks mutually collide, and therefore, they can move one to another, the existence of the friction acting on these contact boundaries will be also assumed. These movements can be also evoked by earthquakes. In this study we will assume the existence of the non-penetrability conditions (known as the Signorini type conditions) and the Coulomb friction acting on these contact boundaries. These non-penetrability conditions describe the facts (i) that the relative normal displacements on the common contact part of the contact boundaries cannot be larger than the distance d between the bodies (lithospheric plates and blocks) being in mutual contact; (ii) that by contact only compressive normal forces can be transmitted; (iii) that normal forces can be transmitted only if there is contact, that is, if $d = 0$. The existence of the friction at the contact boundaries, which is described by the Coulomb law of friction, is possible only if colliding bodies are in mutual contact, i.e., if $d = 0$. The frictional forces acting on the contact boundaries are, in their absolute value, proportional to the normal stress component, where the coefficient of proportionality is the coefficient of Coulomb friction. Then, due to the tangential and frictional forces acting at the contact boundary, we have two cases: (i) if the absolute value of tangential forces is less than the frictional forces, then the frictional forces preclude the mutual shifts of both bodies being in contact and thus in the collision area stresses are further accumulated; (ii) if the tangential forces are equal in their absolute value to the frictional forces, then there are no forces that can preclude the mutual movements of both elastic bodies. Thus the contact points change their position in the direction opposite to that in which the tangential stress components act. This is one mechanism of the earthquake origin as well as the mechanism of the possible aftershocks.

Foreshocks are earthquakes which precede larger earthquakes in the same location. Aftershocks are smaller earthquakes which occur in the same area during the days to years following a larger event - the "main-shock". As a general rule, aftershocks represent minor re-adjustments along the portion of a fault that slipped at the time of the main shock. The frequency of these aftershocks decreases with time.

Studies of rupture processes during earthquakes are performed on the assumption that only shear slip occurs, i.e., only Modes II and III are taken into account. Atkinson (1987) showed that the rupture process of an earthquake is the superposition of the Modes I, II and III of dynamic fracture mechanics. Nevertheless, Modes II and III constitute the most important mechanism for generating seismic energy, ground motions, and shaking. In shallow earthquakes the Mode I is a very important factor. Shear ruptures propagate only along a weaken zone and involve sliding with friction. Rupture propagation is generally modelled using a fracture mechanics approach, likening the rupture to a propagating mixed mode shear crack. The rupture velocity is a function of the fracture energy in the volume around the fracture tip, increasing with decreasing fracture energy. The velocity of rupture propagation is order of magnitude faster than the displacement velocity across the fault. Earthquake ruptures typically propagate at velocities that are in the range $70-90\%$ of the S-wave velocity and that is independent of the earthquake size. A small subset of earthquake ruptures appear to have propagated at speeds greater than the S-wave velocity. These super-shear earthquakes have all been observed during large strike-slip events. Some earthquake ruptures travel at unusually low velocities and are referred to as slow earthquakes. A particularly dangerous form of slow earthquake is the tsunamiearthquake. The definition of the "tsunamiearthquake" was introduced in 1972 by Kanamori as an "earthquake that produces a large size tsunami relative to the value of its surface magnitude (M_S)." Tsunamiearthquakes are characerized by slow rupture propagation, which leads to a lower energy release than expected for the respective seismic moment. The discrepancy between the energy and the moment is responsible for the difference between M_W and M_S.

2.2. Mathematical Models Based on the Nonlinear Thermo-visco-elastic and on the Thermo-visco-plastic Rheologies

In this section we will investigate generation of the earthquake, accompanied with ensuing submarine landslide, mudslide and with the presence of the typhoon, based on the dynamic multibody contact problems with Coulomb friction in the non-linear thermo-visco-elasticity with short memory and damping in N-dimensions. Let $I=(0,t_p)$, $t_p>0$, be a time interval. Let $\Omega \subset \mathbb{R}^N$, $N=2,3$, be a region occupied by a system of r geological bodies[1] of arbitrary shapes Ω^ι such that $\Omega=\cup_{\iota=1}^r \Omega^\iota$. We will assume that all Ω^ι have Lipschitz boundaries $\partial\Omega^\iota$ and that $\partial\Omega=\Gamma_\tau\cup\Gamma_u\cup\Gamma_c$, where the disjoint parts Γ_τ, Γ_u, Γ_c are open subsets. Assume that a system of geological bodies occupying the region Ω consists of r geological bodies Ω^ι, $\iota=1,\ldots,r$, and let several neighboring geological bodies, say Ω^s and Ω^m, are in a mutual contact. Denote the common contact boundary between both lithospheric plates or lithospheric blocks Ω^s and Ω^m before deformation by Γ_c^{sm}. Let $\mathbf{n}=(n_i)$ denote the unit outer normal vector and $\mathbf{t}=(t_i)$ denote the unit tangential vector to the boundary $\partial\Omega$ including the contact boundary Γ_c^{sm}. Further, denote by $\mathbf{u}=(u_i)$ the displacement vector, by $\mathbf{v}=(v_i)=(u_i')$ the velocity vector and by $\boldsymbol{\tau}=(\tau_i)$, $\tau_i=\tau_{ij}n_j$, $i,j=1,\ldots,N$, the stress vector, its normal and tangential components by $\tau_n=\tau_i n_i=\tau_{ij}n_i n_j$, $\boldsymbol{\tau}_t=\boldsymbol{\tau}-\tau_n\mathbf{n}$, and the normal and tangential components of displacement vectors by $u_n=u_i n_i$ and $\mathbf{u}_t=\mathbf{u}-u_n\mathbf{n}$ and by $v_n=v_i n_i$ and $\mathbf{v}_t=\mathbf{v}-v_n\mathbf{n}$ the normal and tangential components of velocity vector. Denote by $\mathbf{u}^s,\mathbf{u}^m,\mathbf{v}^s,\mathbf{v}^m,T^s,T^m$ (indices s,m correspond with the neighboring geological bodies being in contact) the displacements, the velocities and the temperatures in the neighboring geological bodies. All these quantities are functions of the spatial coordinates.

[1]By a geological body we will understand a lithospheric plate, lithospheric block or their parts, failured slope, volcano, part of the oceanic plate, etc.

Moreover, we will assume that $\Gamma_u = {}^1\Gamma_u \cup {}^2\Gamma_u \cup {}^3\Gamma_u \cup {}^4\Gamma_u$, $\Gamma_\tau = {}^1\Gamma_\tau \cup {}^2\Gamma_\tau$ and $\Gamma_c = \cup_{s,m}\Gamma_c^{sm}$, $\Gamma_c^{sm} = \partial\Omega^s \cap \partial\Omega^m$, $s \neq m$, $s,m \in \{1,\dots,r\}$. We denote by $\Omega_t = I \times \Omega$ the time-space domain and by $\Gamma_\tau(t) = \Gamma_\tau \times I$, $\Gamma_u(t) = \Gamma_u \times I$, $\Gamma_c(t) = \Gamma_c \times I$ the parts of its boundary $\partial\Omega_t = \partial\Omega \times I$.

Rheologies

To develop the N-dimensional stress-strain relation in the **non-linear thermo-visco-elastic rheology** $(N = 2,3)$ the strain energy density function $E^{(n)}(e_{ij})$, $n = e,v$, will be used. Let $A^{(n)}$, $n = e,v$, be a scalar-valued function of strains e_{ij}, defined by

$$A^{(n)} = c_{ijkl}^{(n)}(\mathbf{x})e_{ij}(\mathbf{u})e_{kl}(\mathbf{u}), \quad n = e,v, \quad i,j,k,l = 1,\dots,N, \tag{5}$$

where $e_{ij}(\mathbf{u}) = \frac{1}{2}\left(\frac{\partial u_i}{\partial x_j} + \frac{\partial u_j}{\partial x_i}\right)$ are components of the small strain tensor, $c_{ijkl}^{(n)}(\mathbf{x})$ are elastic coefficients for $n = e$ and viscous coefficients for $n = v$, satisfying

$$\begin{aligned}
&c_{ijkl}^{(n)}(\mathbf{x}) \in L^\infty(\Omega), \quad c_{ijkl}^{(n)}(\mathbf{x}) = c_{jikl}^{(n)}(\mathbf{x}) = c_{klij}^{(n)}(\mathbf{x}) = c_{ijlk}^{(n)}(\mathbf{x}),\ n = e,v,\\
&c_{ijkl}^{(n)}(\mathbf{x})e_{ij}e_{kl} \geq c_0^{(n)}e_{ij}e_{ij} \quad \forall e_{ij}, \quad e_{ij} = e_{ji} \text{ and a.e. } \mathbf{x} \in \Omega, \quad c_0^{(n)} > 0 \\
&\qquad\qquad\text{for the anisotropic rock materials,,}\\
&c_{ijkl}^{(n)} = \lambda^{(n)}\delta_{ij}\delta_{kl} + \mu^{(n)}(\delta_{ik}\delta_{jl} + \delta_{il}\delta_{jk}) \text{ for the isotropic rock materials,}
\end{aligned} \tag{6}$$

where $\lambda^{(n)}, \mu^{(n)}$ for $n = e$ are the Lamé elastic coefficients and for $n = v$ the viscous coefficients both for isotropic materials. A repeated index implies summation from 1 to N. In more generalized case we can assume that the elastic and viscous coefficients are nonlinear functions of $\mathbf{u}$.

The stress-strain relation can be derived by using the strain energy density function $E^{(n)} = A^{(n)\lambda}(e_{ij}(\mathbf{u}))$ from the relation

$$\tau_{ij}^{(n)}(\mathbf{u}) = \frac{\partial E^{(n)}(e_{ij}(\mathbf{u}))}{\partial e_{ij}} = \frac{\partial A^{(n)\lambda}(e_{ij}(\mathbf{u}))}{\partial e_{ij}} = c_{ijkl}^{(n)*}e_{kl}(\mathbf{u}),\ n = e,v, \tag{7}$$

where $\tau_{ij}^{(n)}$ are the elastic stresses for $n = e$ and viscous stresses for $n = v$, and where $c_{ijkl}^{(n)*} = 2\lambda\left[A^{(n)}(e_{ij}(\mathbf{u}))\right]^{\lambda-1}c_{ijkl}^{(n)}(\mathbf{x})$ are the strain-dependent non-linear elastic coefficients, λ a positive parameter. The scalar coefficient $2\lambda\left[A^{(n)}(e_{ij}(\mathbf{u}))\right]^{\lambda-1}$ depends on the state of strain. The parameter λ simulates the degree of non-linear properties of materials. The parameter λ characterizes for $\lambda < 1$ softened rock materials, for $\lambda = 1$ elastic rock materials and for $\lambda > 1$ hardened rock materials. The coefficients $c_{ijkl}^{(n)*}$ satisfy the following conditions

$$\begin{aligned}
&c_{ijkl}^{(n)*} = c_{jikl}^{(n)*} = c_{klij}^{(n)*} = c_{ijlk}^{(n)*},\ i,j,k,l = 1,\cdots,N,\ n = e,v,\\
&c_{ijkl}^{(n)*}e_{ij}e_{kl} \geq c_0^{(n)*}e_{ij}e_{ij}, \quad \forall e_{ij}, \quad e_{ij} = e_{ji} \text{ and a.e. } \mathbf{x} \in \Omega,\ c_0^{(n)*} > 0,
\end{aligned} \tag{8}$$

the conditions of symmetry and the Lipschitz conditions.

Since in a liquid $\mu^{(e)} = 0$, then in a liquid P waves propagate only, and therefore, this type of rheology will be useful for our further modeling.

The thermal stesses $\tau_{ij}^{(T)}$, that are generated by a heat, are defined by

$$\tau_{ij}^{(T)} = -\beta_{ij}(T - T_0), \tag{9}$$
$$\beta_{ij} \in C^1(\overline{\Omega}), \quad \beta_{ij} = \beta_{ji}, \, \beta_{ij} \geq \beta_0 > 0, \, i,j = 1,\ldots,N,$$

where β_{ij} and $\beta_{ij} = \gamma\delta_{ij}$ are the coefficients of thermal expansion for anisotropic or for isotropic rocks, respectively, and T and T_0 are the actual and the initial temperatures. A repeated index implies summation from 1 to N.

Then the stress-strain relation in the non-linear thermo-visco-elastic rheology is given by

$$\tau_{ij} = \tau_{ij}^{(e)} + \tau_{ij}^{(v)} + \tau_{ij}^{(T)} = c_{ijkl}^{(e)*} e_{kl}(\mathbf{u}) + c_{ijkl}^{(v)*} e_{kl}(\mathbf{u}') - \beta_{ij}(T - T_0). \tag{10}$$

To develop the constitutive relation of the **(thermo-)visco-plastic rheology** the Bingham type of visco-plastic rheology will be used. The Bingham rheology has the following property: the rock material starts to flow if and only if the applied forces exceed a certain limit, the so-called yield limit. We speak also about Bingham solid/fluid media (Oldroyd (1947), Glowinski et al. (1976), (1981), Nedoma (2006a)).

Let us denote by $\mathbf{u} = (u_i)$ the material velocity, by $D = (D_{ij})$ the strain rate tensor and $D^D = (D_{ij}^D)$ the strain rate deviator defined by

$$D_{ij} = D_{ij}(\mathbf{u}) = \frac{1}{2}\left(\frac{\partial u_i}{\partial x_j} + \frac{\partial u_j}{\partial x_i}\right), \; D_{ij}^D = D_{ij} - \frac{1}{3}D_{kk}\delta_{ij}, \; i,j = 1,\ldots,N, \tag{11}$$

where δ_{ij} is the Kronecker symbol. The invariant of the strain rate tensor is defined as $D_{II} = \frac{1}{2}D_{ij}D_{ij} = \frac{1}{2}|D|^2$, where $|D| = |D(\mathbf{v})| = (\sum_{1\leq i,j\leq N} |D_{ij}(\mathbf{v})|^2)^{\frac{1}{2}}$ is the Frobenius norm of the tensor $D(\mathbf{v})$. For derivation of the stress-strain relation in the Bingham rheology let us denote by $\tau^{(B)} = (\tau_{ij}^{(B)})$ the Cauchy stress tensor and its deviator $\tau^{(B)D} = (\tau_{ij}^{(B)D})$ by

$$\tau_{ij}^{(B)D} = \tau_{ij}^{(B)} - \frac{1}{3}\tau_{kk}^{(B)}\delta_{ij}, \quad \text{i.e.,} \quad \tau^{(B)D} = \tau^{(B)} + pI_N, \, N = 2,3, \tag{12}$$

where $-p$ denotes the spherical part of the stress tensor and has a meaning of the pressure and I_N, $(N = 2,3)$, is the identity tensor. Besides the deviators D_{ij}^D and $\tau_{ij}^{(B)D}$ another tensor deviator $S^D = (S_{ij}^D)$ can be introduced, which corresponds to the plastic properties of the rock material.

For any process in the Bingham rheology we have

$$\tau^{(B)D} = S^D + 2\hat{\mu}D^D, \tag{13}$$

and

$$f(S^D) = \frac{1}{2}|S^D|^2 - \hat{g}^2 \leq 0, \tag{14}$$

that is, the von Mises relation or the yield condition, and

$$D^D = 2\vartheta S^D, \tag{15}$$

where $|S^D|^2 = S^D_{ij}S^D_{ij}$, $\hat{\mu} > 0$ is the threshold of viscosity, $\hat{g}$ is the threshold of plasticity or the yield limit, $\hat{g}/\sqrt{2}$ is the yield stress in pure shear and ϑ is a scalar function defined as (see Nedoma (2006))

$$\vartheta = (4\hat{\mu})^{-1}\left((\tau^{(B)}_{II})^{1/2}\hat{g}^{-1} - 1\right),$$

where $\tau^{(B)}_{II} = \frac{1}{2}\tau^{(B)D}_{ij}\ \tau^{(B)D}_{ij} = \frac{1}{2}|\ \tau^{(B)D}|^2$ is an invariant of the stress tensor in the Bingham rheology.

The von Mises relation $\frac{1}{2}|S^D|^2 - \hat{g}^2 \leq 0$ yields that the invariant $S_{II} = \frac{1}{2}S^D_{ij}S^D_{ij} = \frac{1}{2}|S^D|^2$ cannot reach the square of the yield limit; the relation $S^{1/2}_{II} = 2^{-1/2}|S^D| = \hat{g}$ is called the yield condition.

The constitutive law in the Bingham rheology can be written as

$$\begin{aligned} D_{ij} &= (2\hat{\mu})^{-1}\left(1 - 2^{\frac{1}{2}}\hat{g}|\ \tau^{(B)D}|^{-1}\right)\tau^{(B)D}_{ij} && \text{if} \quad 2^{-1/2}|\ \tau^{(B)D}| > \hat{g}, \\ D_{ij} &= 0 && \text{if} \quad 2^{-1/2}|\ \tau^{(B)D}| \leq \hat{g}, \end{aligned} \tag{16}$$

where $\frac{1}{2}|\ \tau^{(B)D}|^2 = \ \tau^B_{II}$, $\tau^B_{II} = \frac{1}{2}\ \tau^{(B)D}_{ij}\ \tau^{(B)D}_{ij}$ is an invariant of the stress tensor. Note, that in some papers the constitutive relation is introduced in the deviatoric formulation.

To invert relations (16) we have: let $|D| = 0$, then from (16) we find $2^{-1/2}|\ \tau^{(B)D}| \leq \hat{g}$; if $|D| \neq 0$ then from (16) we obtain $2^{-1/2}|\ \tau^{(B)D}| > \hat{g}$ and $|\ \tau^{(B)D}| = 2\hat{\mu}|D| + 2^{\frac{1}{2}}\hat{g}$. Then using the incompressibility condition $D_{kk} = 0$ and (11) from (16) we find

$$\tau^{(B)}_{ij} = -p\delta_{ij} + \hat{g}D_{ij}D^{-1/2}_{II} + 2\hat{\mu}D_{ij}, \quad \text{provided} \quad D_{II} \neq 0, \tag{17}$$

representing the constitutive stress-strain rate relation in the Bingham rheology. For $\hat{g} = 0$, we have a classical viscous incompressible (Newtonian) fluid; for small $\hat{g} > 0$, we have strongly visco-plastic materials close to the classical viscous fluid (e.g. muddy flows), for $\hat{g}$ becomes sufficiently large, the fluid stops flowing (we speak about the blocking property, see Glowinski et al. (1976)), and for $\hat{g} \to \infty$, we have absolutely rigid rock materials. Therefore, if $\hat{g}$ is strictly positive, we can observe rigid zones inside the flow; when $\hat{g}$ further increases, these rigid zones become larger and larger (i.e., it models e.g. the sediments) and start completely rigid when $\hat{g}$ is sufficiently large. In this case, when these rigid zones block the flow, we speak about blocking property of the Bingham fluid, that in the submarine landslide modeling ensures stability of the submarine slopes. In the rigid zones $D(\mathbf{u}) = 0$. About the Bingham rheology some of authors speak as about the Bingham solid (see e.g. Oldroyd (1947)). It is evident that this Bingham rheology is also useful for the submarine moody flows and submarine landslides of soil and rock masses saturated by water as well as for the oceanic water.

The stress-strain rate relation in thermo-Bingham rheology is as follows:

$$\begin{aligned} \tau_{ij} = \tau^{(B)}_{ij} + \tau^{(T)}_{ij} &= -p\delta_{ij} + \hat{g}D_{ij}D^{-1/2}_{II} + 2\hat{\mu}D_{ij} - \beta_{ij}(T - T_0) = \\ &= -p\delta_{ij} + 2^{\frac{1}{2}}\hat{g}D_{ij}|D|^{-1/2} + 2\hat{\mu}D_{ij} - \beta_{ij}(T - T_0). \end{aligned} \tag{18}$$

Equation of motion and equation of heat conduction

Let us assume that $\mathbf{F}(t,\mathbf{x}),\mathbf{P}(t,\mathbf{x})$ be the acting body and surface forces, $\rho(\mathbf{x})$ be the density satisfying the conservation of mass through the continuity equation

$$\frac{\partial\rho}{\partial t}+\frac{\partial\rho u_i}{\partial x_i}=0 \quad \text{for compressible rock materials,}$$
$$\frac{\partial u_i}{\partial x_i}=0 \qquad\qquad \text{for incompressible rock materials,}$$

W be the thermal sources, $c_e(\mathbf{x})$ be the specific heat, $c(\mathbf{x})=\rho(\mathbf{x})c_e(\mathbf{x})$ the thermal capacity and β_{ij} the coefficient of linear thermal expansion. Let us denote by "$'$" the respective time derivatives and let us denote by $\mathbf{u}=(u_i)$ the displacement vector and by $\mathbf{u}'=(u_i')$ the velocity vector. Let on Γ_c^{sm} the positive direction of the outer normal vector $\mathbf{n}$ be assumed with respect to Ω^s. Further, we denote by $[u_n]^{sm}=u_n^s-u_n^m$ the jump of the normal component of the displacement vector across the contact zone between neighboring geological bodies - lithospheric plates and lithospheric blocks in the subduction zone, Ω^s and Ω^m and similarly we put $[\mathbf{u}_t]^{sm}=\mathbf{u}_t^s-\mathbf{u}_t^m$ for the tangential component of the displacement vector. Furthermore, $T(t,\mathbf{x})$, $T_0(\mathbf{x})$ are the actual and initial temperatures and $[T]^{sm}=T^s-T^m$. Note, that if one of the colliding bodies is absolutely rigid, say Ω^m, then $[u_n]^{sm}\equiv u_n^s$, and similarly $[\mathbf{u}_t]^{sm}=\mathbf{u}_t^s$, $[T]^{sm}=T^s$.

The seismic waves propagate from their origins situated in the subduction zone to all the sites of the Earth's body. Therefore, for our models we need to limit ourselves to the finite (3D) region, that covers up the subduction zone, a part of the continent and a part of the ocean (Fig.1). From this point of view, we introduce the function $\alpha(\mathbf{x})\geq 0$, that physically means the damping, and such that in certain sufficiently large distances from the earthquake origin, say $|X|$, it will depend on the seismic wave lengths and such that the propagated seismic waves in distances greater than $|X|$ are fully damped, whereas in the investigated region (i.e., for $x<|X|$) $\alpha(\mathbf{x})=\alpha_\mathbf{0}(\mathbf{x})$, where α_0 corresponds to the natural rock damping. By this mentioned way then from the Earth's body we determine the finite region for our mathematical modeling of propagations of seismic waves and of the ensuing tsunami and the effects of evoked submarine landslides and mudslides as well as of effects of typhoon, that propagates at the same time in the determined finite region around the subduction zone. The determined area will be denoted by Ω and we will assume that it is created by $r\geq 2$ subregions, where $\Omega=\cup_{\iota=1}^r\Omega^\iota$ and where $\Omega^\iota,\iota=1,\cdots,r$, denotes the separated lithospheric plates and lithospheric blocks from the region of our interest.

From the momentum conservation law and the entropy density law the equation of motion and the expanded equation of heat conduction (heat equation) are as follows

$$\rho^\iota\left(\frac{\partial^2 u_i^\iota}{\partial t^2}+u_j'\frac{\partial u_i}{\partial t}\right)+\alpha^\iota\frac{\partial u_i^\iota}{\partial t}+2\rho\varepsilon_{ijk}\omega_j u_k'=\frac{\partial\tau_{ij}^\iota(\mathbf{u}^\iota,\mathbf{u}^{\iota\prime},T^\iota)}{\partial x_j}+F_i^\iota,\ \text{(a)} \qquad (19)$$

$$\rho^\iota c_e^\iota\frac{\partial T^\iota}{\partial t}-\frac{\partial}{\partial x_i}(\kappa_{ij}^\iota\frac{\partial T^\iota}{\partial x_j})=W^\iota-\rho^\iota\beta_{ij}^\iota T_0^\iota e_{ij}(\mathbf{u}^{\iota\prime})$$
$$+c_{ijkl}^{(v)*\iota}e_{kl}(\mathbf{u}^{\iota\prime})e_{ij}(\mathbf{u}^{\iota\prime}), \qquad \text{(b)}$$

where $i,j=1,\ldots,N$, $(t,\mathbf{x})\in\Omega_t^\iota=I\times\Omega^\iota$, $\iota=1,\ldots,r$,

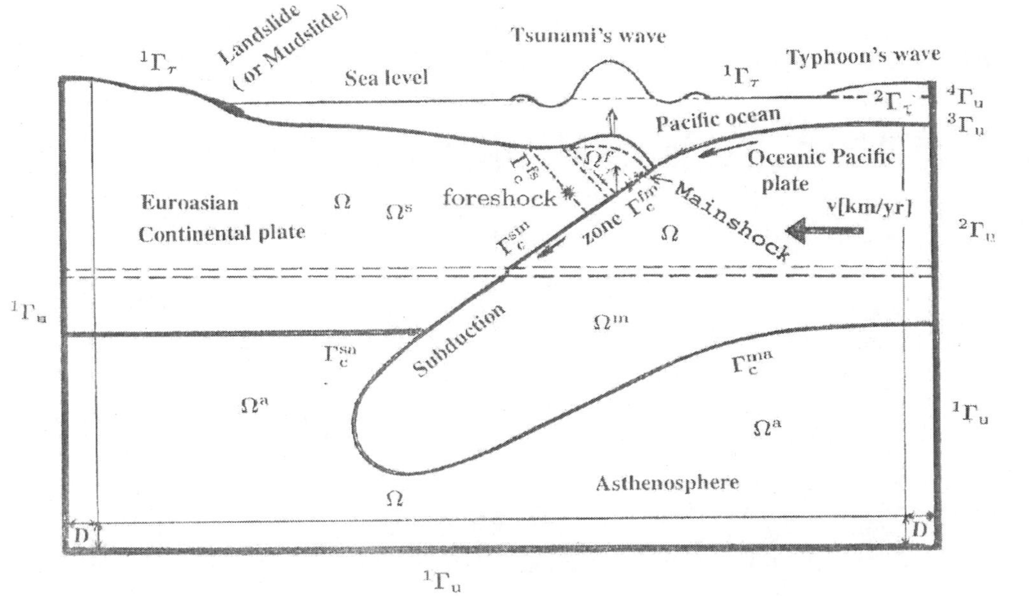

Figure 1. Schematic cross-section of the subduction zone and the corresponding mathematical model of mechanisms of the tsunamiearthquakees in areas with typhoons.

where τ_{ij} is the stress tensor, defined by the stress-strain relations introduced above, and where $\mathbf{u}$, and $\mathbf{u}'$ are the displacement and velocity vectors, T and T_0 are the actual and initial temperatures of non-deformable parts of the system of lithospheric plates and lithospheric blocks in the area of the subduction zone created the investigated region Ω, $\kappa^{\iota}_{ij}(\mathbf{x})$, $\iota = 1,\ldots,r$, are the non-linear coefficients of thermal conductivity depending on the space variable, the temperature and the gradient of temperature, that are continuous in the sense of Carathéodory (i.e., $\mathbf{x} \mapsto \kappa^{\iota}_{ij}(\mathbf{x}, T^{\iota}, \nabla T^{\iota})$ is measurable for every $T^{\iota} \in \mathbb{R}$ and $T^{\iota} \mapsto \kappa^{\iota}_{ij}(\mathbf{x}, T^{\iota}, \nabla T^{\iota})$ is continuous for almost every $\mathbf{x} \in \Omega^{\iota}$) and satisfy

$$\begin{aligned}
&\kappa^{\iota}_{ij}(\mathbf{x}, T^{\iota}, \nabla T^{\iota}) = \kappa^{\iota}_{ji}(\mathbf{x}, T^{\iota}, \nabla T^{\iota}), \\
&\kappa^{\iota}_0(1+\delta_0 \mid T^{\iota} \mid^{\gamma_0} + \delta_1 \mid \nabla T^{\iota} \mid^{\gamma_1})\zeta_i\zeta_j \le \kappa^{\iota}_{ij}(\mathbf{x}, T^{\iota}, \nabla T^{\iota})) \le \\
&\le \kappa^{\iota}_1(1+\delta_0 \mid T^{\iota} \mid^{\gamma_0} + \delta_1 \mid \nabla T^{\iota} \mid^{\gamma_1})\zeta_i\zeta_j \\
&\text{for all } \{\zeta_i\} \in \mathbb{R}^N,\ N = 2,3, \text{ and all } \mathbf{x} \in \Omega \text{ with } 0 < \kappa^{\iota}_0 \le \kappa^{\iota}_1 < +\infty, \\
&\text{and with } \gamma_0 > 0,\ \gamma_1 > 0, \delta_0 > 0,\ \delta_1 > 0, \iota = 1,\cdots,r,
\end{aligned} \tag{20}$$

and where the term $\rho^{\iota}\beta^{\iota}_{ij}T^{\iota}_0 e_{ij}(\mathbf{u}^{\iota\prime}), \iota = 1,\cdots,r$, represents the deformation energy that changes into a heat, the so-called deformation heat and the term $c^{(v)*\iota}_{ijkl}\, e_{ij}(\mathbf{u}^{\iota\prime})\, e_{kl}(\mathbf{u}^{\iota\prime})$ is the additional viscous dissipation and represents one of heat sources in the heat equation, the so-called viscous heat, that satisfies the first thermodynamic law, and where $c^{(n)*}_{ijkl}$, $n = e, v$, satisfy the conditions of symmetry and of the Lipschitz continuity, defined above. Moreover, β_{ij} denotes the coefficient of thermal expansion, F_i denote the body forces, $e_{ij}(\mathbf{u})$ is the small strain tensor, $e_{ij}(\mathbf{u}')$ is small strain rate tensor. The terms $\alpha^\iota \frac{\partial u_i^\iota}{\partial t}$ and $2\rho\varepsilon_{ijk}\omega_j u_k'$ denote the effects of damping in the first case and the effect of the rotation of the Earth in the second case, and where $\alpha = \alpha_0(\mathbf{x}) \geq 0$ represents the non-homogeneous elastic wave damping in the investigated region Ω and $\alpha = \alpha_0(\mathbf{x}) + \alpha_D(\mathbf{x}, \lambda^s, D)$ at the moment when the frictional forces are not sufficient to preclude the mutual shifting of both geological bodies - the lithospheric plates or the lithospheric blocks in the subduction zone (i.e., at the moment of the earthquake origin), and such that the seismic waves near the boundary Γ_u propagated through the distance D from the boundary Γ_u (i.e., for distances from the earthquake origin $\mathbf{x} > |\mathbf{X}|$) are fully damped, i.e., any reflected waves do not return back into the investigated domain Ω, λ^s denotes the corresponding smallest wave length (for details see Nedoma (1998)), ω_j is the component of the Earth's angular velocity, ε_{ijk} is the third order tensor (Levi-Civita's tensor) and $\varepsilon_{ijk} = 1$ if i, j, k are in cyclic order, $\varepsilon_{ijk} = -1$ if i, j, k are in anti-cyclic order, and $\varepsilon_{ijk} = 0$ if any pair of indices or all three indices have the same value.

Boundary and contact conditions

Let us assume that the system of geological bodies, the lithospheric plates and lithospheric blocks, are loaded, and moreover, that their parts in the subduction zone are in mutual contacts. Further, let us assume that the volume forces and surface tractions change in time and the movements of geological plates and lithospheric blocks change in time. Then we introduce the contact conditions in displacements (or in velocities) and also the conditions for the temperature.

Let the boundary Γ_τ, the Earth's surface, be divided in two parts, say ${}^1\Gamma_\tau$ and ${}^2\Gamma_\tau$, such that $\Gamma_\tau = {}^1\Gamma_\tau \cup {}^2\Gamma_\tau$. Let us assume that a portion of the studies system of lithospheric plates and lithospheric blocks on one part of this boundary is loaded, we denote it as ${}^2\Gamma_\tau$, while on the second one, the surface of the Earth and the sea level, we denote it as ${}^1\Gamma_\tau$, is unloaded. Let the temperature $T_1(\mathbf{x}, \mathrm{t})$ be given. Then we will assume the condition in the form

$$
\begin{aligned}
&T(\mathbf{x},t) = T_1(\mathbf{x},t), \quad \tau_{ij}n_j = P_i, \quad i,j = 1,\dots,N,, \\
&\text{for } (t,\mathbf{x}) \in {}^l\Gamma_\tau(t) = I \times \cup_{\iota=1}^{r}({}^l\Gamma_\tau \cap \partial\Omega^\iota), \quad \text{meas}\, {}^l\Gamma_\tau \geq 0, \ l = 1,2,
\end{aligned}
\tag{21}
$$

where $P_i = 0$ or $P_i \neq 0$ and $T_1 = 0$ or $T_1 > 0$ on ${}^l\Gamma_\tau$, $l = 1,2$.

Let us assume that the boundary Γ_u be divided in two parts, say ${}^1\Gamma_u$ and ${}^2\Gamma_u$, so that $\Gamma_u = {}^1\Gamma_u \cup {}^2\Gamma_u$. Let the non-zero external temperature Y and the displacement vector $\mathbf{u_2}$ be given at a part of the boundary Γ_u, we denote it by ${}^1\Gamma_u$, that due to a damping $\mathbf{u_2} = \mathbf{0}$. Thus, we will assume e.g. the conditions

$$
\begin{aligned}
&\kappa_{ij}\frac{\partial T(\mathbf{x},t)}{\partial x_j} n_i = K(Y(\mathbf{x}) - T(\mathbf{x},t)), \quad u_i = 0, \quad i = 1,\dots,N, \\
&\text{for } (t,\mathbf{x}) \in {}^1\Gamma_u(t) = I \times \cup_{\iota=1}^{r}({}^1\Gamma_u \cap \partial\Omega^\iota), \quad \text{meas}\, {}^1\Gamma_u > 0.
\end{aligned}
\tag{22}
$$

Let us assume that on a portion of the system of lithospheric plates and lithospheric blocks on a certain part of the boundary ${}^2\Gamma_u$ the displacements $\mathbf{u_2}$ is non-zero, that models on the individual parts of the boundary ${}^2\Gamma_u$ (where for simplicity we denoted ${}^2\Gamma_u \equiv {}^2\Gamma_u \cup {}^3\Gamma_u \cup {}^4\Gamma_u$) the movements of the subducted oceanic (Pacific) plate (i.e., on ${}^2\Gamma_u$), the motion of the oceanic water (i.e., on ${}^3\Gamma_u$) and the motion of water generated by the typhoon (i.e., on ${}^4\Gamma_u$), but that the heat flow is non-zero at this part of the boundary ${}^2\Gamma_u$, that is, a heat flow streams into the investigated region from the neighborhoods, i.e.,

$$\kappa_{ij}\frac{\partial T(\mathbf{x},t)}{\partial x_j}n_i = q, \quad u_i = u_{2i}, \quad i = 1,\ldots,N, \tag{23}$$

$$\text{for } (t,\mathbf{x}) \in {}^2\Gamma_u(t) = I \times \cup_{\iota=1}^{r}({}^2\Gamma_u \cap \partial\Omega^\iota), \quad \text{meas } {}^2\Gamma_u > 0.$$

To describe the contact between lithospheric plates and lithospheric blocks in the subduction zone we will consider the normal approach and the tangential process. By **contact conditions** we understand conditions that describe the tangential process, that is a relation involving the tangential stress $\boldsymbol{\tau}_t$ and the tangential displacements $\mathbf{u}_t$ (or the tangential velocity $\mathbf{v}_t = \mathbf{u}_t'$, where we denote $\mathbf{u}' = \frac{\partial \mathbf{u}}{\partial t}$). The nature of friction forces acting between lithospheric plates and lithospheric blocks in the subduction zone is extremely complex and depends on several factors, like as the constitution of the interface, the time scales, the response of the interface to normal forces, the presence of melting, inertia and thermal effects, roughness of the contacting surfaces, history of loadings, etc. Such effects and properties of contact surfaces can be formulated as nonlinear equalities or inequalities.

Let us introduce the normal and tangential components of the displacement and velocity vectors and the stress vector by $u_n = \mathbf{u}.\mathbf{n}$, $\mathbf{u}_t = \mathbf{u} - u_n\mathbf{n}$, $v_n = \mathbf{v}.\mathbf{n}$, $\mathbf{v}_t = \mathbf{v} - v_n\mathbf{n}$, $\boldsymbol{\tau} = (\tau_{ij}n_j)$, $\tau_n = \tau_{ij}n_jn_i$, $\boldsymbol{\tau}_t = \boldsymbol{\tau} - \tau_n\mathbf{n}$. The component $\boldsymbol{\tau}_t$ is connected with friction forces on the contact surface $\cup\Gamma_c^{sm}$.

Because in the subduction zone the volcanic activity is also indicated and the volcanic chambers are observed, then such conditions must described also the consequences connected with the volcanism in this region. The conditions on the contact boundaries are represented by the non-penetration (Signorini) conditions coupled with the Coulomb law of friction as well as by conditions describing geothermal processes in the subduction zone. Similarly, these conditions must describe the processes connected with landslides and mudslides in this region. Moreover, these contact conditions will simulate the consequences of strongly loaded lithospheric plates and lithospheric blocks created the subduction zone. These conditions are represented by the **unilateral contact conditions of Signorini types** with friction or without friction in the case if the contact surface between lithospheric plates and lithospheric blocks are partially melted, and therefore, the coefficients of friction rapidly decrease. In such cases during the deformation process the contact points of the colliding lithospheric plates and lithospheric blocks will displace in different ways, but colliding parts of the lithospheric plates and lithospheric blocks can not penetrate, i.e.,

$$[u_n]^{sm} \le d^{sm} \quad \text{in displacements or} \quad [v_n]^{sm} \le 0 \quad \text{in velocities on } \Gamma_c^{sm} \times I, \tag{24}$$

where the positive direction of the outward normal to the contact boundary, denoted as $\mathbf{n}$, is related to the domain Ω^s occupying by the s-th part of the investigated subduction zone, and where d^{sm} denotes the gap between the neighbouring colliding lithospheric plates and lithospheric blocks. This relation is known as the condition of non-penetration.

For contact forces from the law of action and reaction it follows that $\tau_n^s = \tau_n^m = \tau_n^{sm}$ and $\boldsymbol{\tau}_t^s = -\boldsymbol{\tau}_t^m = \boldsymbol{\tau}_t^{sm}$. Since normal components cannot be positive (tractions) then

$$\tau_n^{sm} \leq 0 \quad \text{on} \quad \Gamma_c^{sm} \times I. \tag{25}$$

During the deformation of both neighboring colliding lithospheric plates and lithospheric blocks (i) are in a contact, i.e., $[u_n]^{sm} = d^{sm}$ in displacements or $[v_n]^{sn} = 0$ in velocities or (ii) are not in a contact, i.e., $[u_n]^{sm} < d^{sm}$ in displacements or $[v_n]^{sm} < 0$ in velocities. If the lithospheric plates and lithospheric blocks are in contact, i.e., $[u_n]^{sm} = d^{sm}$ in displacement or $[v_n]^{sm} = 0$, then there may exist non-zero contact forces $\tau_n^s = \tau_n^m \equiv \tau_n^{sm} < 0$. If lithospheric plates and lithospheric blocks are not in contact, i.e., if $[u_n]^{sm} < d_{sm}$, then the contact forces are zero, i.e., $\tau_n^s = \tau_n^m \equiv \tau_n^{sm} = 0$. These cases are included in the following condition

$$([u_n]^{sm} - d_{sm})\tau_n^{sm} = 0 \text{ in displacements, } [v_n]^{sm}\tau_n^{sm} = 0$$
$$\text{in velocities on } \Gamma_c^{sm} \times I. \tag{26}$$

Equations (24)-(26) are known as the **Signorini type conditions**. In some real cases gaps d^{sm} between the colliding parts of geological bodies measured along the outward normals, that created cavities, are observed. Then these cavities are filled by the melted rock materials.

Further, if both lithospheric plates or the lithospheric blocks in the subduction zone are in contact, then on the contact boundary, we denote it as Γ_c^{sm}, the Coulomb type of friction acts. The frictional forces $g_c^{sm} = \mathcal{F}_c^{sm}|\tau_n^{sm}|$ acting on the contact boundary Γ_c^{sm} are, in their absolute value, proportional to the normal stress component, where the coefficient of proportionality is the coefficient of Coulomb friction $\mathcal{F}_c^{sm}(\mathbf{u}_t)$ in displacements and $\mathcal{F}_c^{sm}(\mathbf{u}_t')$ in velocities, i.e.,

$$\begin{aligned} g_c^{sm}(\mathbf{x},t) &= \mathcal{F}_c^{sm}(\mathbf{u}_t)|\tau_n^{sm}(\mathbf{x},t)| \quad \text{in displacements,} \\ g_c^{sm}(\mathbf{x},t) &= \mathcal{F}_c^{sm}(\mathbf{u}_t')|\tau_n^{sm}(\mathbf{x},t)| \quad \text{in velocities,} \end{aligned} \tag{27}$$

where $\tau(\mathbf{x},t) = \tau(\mathbf{u},\mathbf{x},t)$ in the (visco-)elastic rheology and $\tau(\mathbf{x},t) = \tau(\mathbf{u},T,\mathbf{x},t)$ in the thermo-visco-elastic rheology, respectively.

Due to the acting and frictional forces we have the following cases:

If the tangential forces $\boldsymbol{\tau}_t^{sm}$ in their absolute values are less than the frictional forces g_c^{sm}, then the frictional forces preclude the mutual shifts of both geological bodies being in contact and the deformation energy is further accumulated. If the tangential forces $\boldsymbol{\tau}_t^{sm}$ in their absolute values reach the frictional forces, so that there are no forces which can preclude the mutual (i.e., bilateral) motions of both geological bodies. Thus the contact points change their positions in the directions opposite to that in which the tangential stress components act. These conditions are described by the following conditions:

$$
\begin{array}{ll}
\text{a)} & \text{if } [u_n]^{sm} = 0 \text{ then } |\boldsymbol{\tau}_t^{sm}(\mathbf{x},t)| \le g_c^{sm}(\mathbf{x},t) \text{ in displacements,} \\
& \text{if } [u'_n]^{sm} = 0 \text{ then } |\boldsymbol{\tau}_t^{sm}(\mathbf{x},t)| \le g_c^{sm}(\mathbf{x},t) \text{ in velocities,} \\
\text{b)} & \text{if } |\boldsymbol{\tau}_t^{sm}(\mathbf{x},t)| < g_c^{sm}(\mathbf{x},t) \text{ then } [\mathbf{u}'_t(\mathbf{x},t)]^{sm} = 0, \\
& \text{which means that the friction forces are sufficient to preclude} \\
& \text{the mutual shifting between the assumed contact bodies and} \\
\text{c)} & \text{if } |\boldsymbol{\tau}_t^{sm}(\mathbf{x},t)| = g_c^{sm} \text{ then there exists a function } \vartheta \ge 0 \text{ such that} \\
& [\mathbf{u}'_t(\mathbf{x},t)]^{sm} = -\vartheta \boldsymbol{\tau}_t^{sm}(\mathbf{x},t), \\
& \text{which means that the friction forces are not sufficient to preclude} \\
& \text{the mutual (bilateral) shifting of both assumed lithospheric plates or} \\
& \text{lithospheric blocks. This shift acts in an opposite direction to} \\
& \text{the acting tangential forces. As introduced above at the} \\
& \text{same moment we put} \\
& \alpha = \alpha_0(\mathbf{x}) + \alpha_D(\mathbf{x}, \lambda^s, D).
\end{array}
\tag{28}
$$

Remark 1. *When $\mathcal{F}_c^{sm} = 0$ then $g_c^{sm} = 0$ and then $\boldsymbol{\tau}_t^{sm} = 0$, and we speak about contact problems without friction.*

On the contact boundary between the lithospheric plates or the lithospheric blocks we will assume that the temperatures and the heat flows are continuous, or moreover, that on the contact boundary the thermal sources created by the frictions, the frictional heats, are generated, i.e.,

$$
T^s = T^m, \kappa_{ij}\frac{\partial T}{\partial x_j} n_{i|(s)} - \kappa_{ij}\frac{\partial T}{\partial x_j} n_{i|(m)} = \mathcal{F}_c^{sm} |\tau_n| \left| [\mathbf{u}'_t]^{sm} \right| \quad \text{on} \quad \Gamma_{c.}^{sm}. \tag{29}
$$

In the case if both colliding plates or lithospheric blocks are in contact then on the contact surface for the case of bilateral contact with Coulomb friction we have the conditions

$$
\begin{array}{l}
[u_n]^{sm} = 0 \quad \text{and} \quad |\boldsymbol{\tau}_t^{sm}| \le \mathcal{F}_c^{sm} |\tau_n^{sm}| \equiv g_c^{sm}, \\
\left\{
\begin{array}{lll}
\text{if} & |\tau_t^{sm}| < g_c^{sm} & \text{then} \quad [\mathbf{u}_t]^{sm} = 0, \\
\text{if} & |\boldsymbol{\tau}_t^{sm}| = g_c^{sm} & \text{then there exists} \quad \lambda \ge 0 \\
& & \text{such that} \quad [\mathbf{u}_t]^{sm} = -\lambda \boldsymbol{\tau}_t^{sm},
\end{array}
\right\} \quad \text{on } \cup_{s,m} \Gamma_c^{sm} \times I,
\end{array}
\tag{30}
$$

in displacements and

$$
\begin{array}{l}
[u'_n]^{sm} = 0 \quad \text{and} \quad |\boldsymbol{\tau}_t^{sm}| \le \mathcal{F}_c^{sm} |\tau_n^{sm}| \equiv g_c^{sm}, \\
\left\{
\begin{array}{lll}
\text{if} & |\boldsymbol{\tau}_t^{sm}| < g_c^{sm} & \text{then} \quad [\mathbf{u}'_t]^{sm} = 0, \\
\text{if} & |\boldsymbol{\tau}_t^{sm}| = g_c^{sm} & \text{then there exists} \quad \lambda \ge 0 \\
& & \text{such that} \quad [\mathbf{u}'_t]^{sm} = -\lambda \boldsymbol{\tau}_t^{sm},
\end{array}
\right\} \quad \text{on } \cup_{s,m} \Gamma_c^{sm} \times I,
\end{array}
\tag{31}
$$

in velocities. The relations (30) and (31) represent the bilateral contact conditions with the Coulomb law of friction that describe the situation when the lithospheric plates or lithospheric blocks in the subduction zone are in mutual contacts.

The amplitude of the Coulomb coefficient of friction is not known, but for the existence of a solution it can be estimated (for the elastic case see Jarušek (1983), Hlaváček et al. (1988) and for thermo-elastic case see Jarušek (1996), Eck et al. (2005), Nedoma (1987), (1998) and in the literature presented here). Then for the isotropic and for the anisotropic cases we have the estimates

$$\| \mathcal{F}_c^{sm} \| < (\mu/(\lambda+2\mu))^{\frac{1}{4}} \sim (c_s/c_p)^{\frac{1}{2}} \quad \text{or} \quad \| \mathcal{F}_c^{sm} \| < a_0/(2A_0), \tag{32}$$

where λ and μ are the Lamé coefficients, the velocities of S and P waves are defined by $c_s^2 = \mu\rho^{-1}$ and by $c_p^2 = (\lambda+2\mu)\rho^{-1}$, a_0 and A_0 are constants, the so-called coefficients of ellipticity and the upper bound of the Hooke tensor, i.e., $a_0\xi_{ij}\xi_{ij} \leq c_{ijkl}\xi_{ij}\xi_{kl} \leq A_0\xi_{ij}\xi_{ij}$ holds. We see that the coefficient of friction depends on the material properties only.

3. Mathematical Model of Earthquakes in Regions with Typhoon Based on the Multibody Contact Problem in Non-linear Thermo-visco-elastic Rheology with Short Memory: Computational Methods and Algorithms

3.1. The Model

In this section we will discuss the model that will facilitate to simulate the earthquake origin and ensuing tsunami, the effect of typhoon as well as the effects of volcanic activity, the submarine landslides and mudslides that may be evoked by the tsunamiearthquake and its aftershocks that can evoke the ensuing further tsunamis. All these effects need to satisfy the main rheological properties in the investigated region of our interest. For detailed studies of landslides and mudslides we refer the reader to the author's study in Nedoma (2010). Therefore, the suitable choice of the rheology seems to be the fundamental for the description of all the effects concerning with the evolution of the tsunamis and of the typhoon's waves and then for the security of the distressed continental areas. It is evident that the presented method will give some estimation about situation in the distressed areas closed the subduction zone, and then after simulations of all the above mentioned processes before their real events it can give the necessary steps to preclude tragical situations in the large distressed continental areas.

In the real Earth it is physically more realistic to describe the problem as the dynamic contact problem in the useful rheology - the non-linear thermo-visco-elastic or the thermo-visco-plastic rheologies. But this bring an additional complications due to the hyperbolic nature of elastodynamics and due to a nature of Coulomb friction because in this case a convexity is lost and there are no good compactness properties. From these point of views mathematical analyses (existence and uniqueness) in continuous domains are limited only for the simple versions of problems.

The coupling terms describe the heat (i) generated by the deformation of the rock bodies and we speak about a deformation heat; (ii) generated by the viscous inner friction (viscosity) and we speak about a viscous heat; (iii) generated by the boundary (contact) friction and we speak about a frictional heat, and moreover, additional terms describing the mutual exchange of energy between the visco-elastic system and the heat equation. Therefore, we then find thermal stresses in the visco-elastic system - in the equation of motion, and a deformation heat in the heat equation.

Let us assume that the subduction zone being of our interest (Fig.1), and denoted as $\Omega \subset \mathbb{R}^N$, $N = 2,3$, is a region occupied by a system of r geological bodies, that is, a system of lithospheric plates and lithospheric blocks of arbitrary shapes Ω^ι, such that $\Omega = \cup_{\iota=1}^r \Omega^\iota$, and let its boundary be denoted as $\partial\Omega$. Let us assume that all Ω^ι have Lipschitz boundaries $\partial\Omega^\iota$ and that $\partial\Omega = \Gamma_\tau \cup \Gamma_u \cup \Gamma_c$, where the disjoint parts Γ_τ, Γ_u, Γ_c are open subsets. Moreover, let us assume that several neighboring geological bodies, say Ω^s and Ω^m, are in a mutual contact. Denote the common contact boundary between both lithospheric plates or lithospheric blocks Ω^s and Ω^m before deformation by Γ_c^{sm}. Let $\mathbf{n} = (n_i)$ denote the unit outer normal vector and $\mathbf{t} = (t_i)$ denote the unit tangential vector to the boundary $\partial\Omega$ including the contact boundary Γ_c^{sm}. Further, denote $\mathbf{u} = (u_i)$ the displacement vector, $\mathbf{v} = (v_i) = (u_i')$ the velocity vector and $\boldsymbol{\tau} = (\tau_i)$, $\tau_i = \tau_{ij}n_j$, $i,j = 1,\ldots,N$, the stress vector, its normal and tangential components by $\tau_n = \tau_i n_i = \tau_{ij}n_i n_j$, $\boldsymbol{\tau}_t = \boldsymbol{\tau} - \tau_n\mathbf{n}$, and the normal and tangential components of displacement vector by $u_n = u_i n_i$ and $\mathbf{u}_t = \mathbf{u} - u_n\mathbf{n}$ and $v_n = v_i n_i$ and $\mathbf{v}_t = \mathbf{v} - v_n\mathbf{n}$ the normal and tangential components of velocity vector. Denote by $\mathbf{u}^k, \mathbf{u}^l, \mathbf{v}^s, \mathbf{v}^m, T^s, T^m$ (indices s,m correspond with the neighboring geological bodies being in contact) the displacements, velocities and the temperatures in the neighboring geological bodies. All these quantities are functions of spatial coordinates and time. Moreover, we will assume that $\Gamma_u = {}^1\Gamma_u \cup {}^2\Gamma_u$, $\Gamma_\tau = {}^1\Gamma_\tau \cup {}^2\Gamma_\tau$ and $\Gamma_c = \cup_{s,m}\Gamma_c^{sm}$, $\Gamma_c^{sm} = \partial\Omega^s \cap \partial\Omega^m$, $s \neq m$, $s,m \in \{1,\ldots,r\}$. Let $I = (0,t_p), t_p > 0$, be the time interval. We denote by $\Omega(t) = I \times \Omega$ the time-space domain and by $\Gamma_\tau(t) = \Gamma_\tau \times I$, $\Gamma_u(t) = \Gamma_u \times I$, $\Gamma_c(t) = \Gamma_c \times I$ the parts of its boundary $\partial\Omega(t) = \partial\Omega \times I$. For simplicity we put as above ${}^2\Gamma_u \equiv {}^2\Gamma_u \cap {}^3\Gamma_u \cup {}^4\Gamma_u$, where by the invidual parts of the boundary ${}^2\Gamma_u$ we can simulate movements of the invasing (subducted) oceanic plate, the motions of the oceanic water and the motion of the water evoked by the typhoon. In our next study we limit ourselves to simpler version of the equation of motin, because in this studied case the second term in the equation of motion and the influence of rotation of the Earth can be neglected.

Then we have to solve the following model problem:

Problem $(\mathcal{P}_{sm})$: Let $N = 2,3$, $r \geq 2$. Find a pair of functions $(\mathbf{u},T) : \Omega \times I \to (\mathbb{R}^N \times \mathbb{R}) \times I$, $N = 2,3$,, satisfying

$$\rho^\iota \frac{\partial^2 u_i^\iota}{\partial t^2} + \alpha^\iota \frac{\partial u_i^\iota}{\partial t} = \frac{\partial \tau_{ij}^\iota(\mathbf{u}^\iota, \mathbf{u}^{\iota\prime}, T^\iota)}{\partial x_j} + F_i^\iota, \tag{33}$$

$$\rho^\iota c_e^\iota \frac{\partial T^\iota}{\partial t} - \frac{\partial}{\partial x_i}(\kappa_{ij}^\iota \frac{\partial T^\iota}{\partial x_j}) = W^\iota - \rho^\iota \beta_{ij}^\iota T_0^\iota e_{ij}(\mathbf{u}^{\iota\prime}) + c_{ijkl}^{(v)\iota *} e_{ij}(\mathbf{u}^{\iota\prime}) e_{kl}(\mathbf{u}^{\iota\prime}), \tag{34}$$

where $i,j = 1,\ldots,N$, $(t,\mathbf{x}) \in \Omega^\iota(t) = I \times \Omega^\iota$, $\iota = 1,\ldots,r$,

with the constitutive stress-strain relation

$$\tau_{ij} = \tau_{ij}^{(e)} + \tau_{ij}^{(v)} + \tau_{ij}^{(T)} = c_{ijkl}^{(e)*}(\mathbf{u})e_{kl}(\mathbf{u}) + c_{ijkl}^{(v)*}(\mathbf{u})e_{kl}(\mathbf{u}') - \beta_{ij}(T - T_0), \tag{35}$$

and the boundary value conditions

$$T(\mathbf{x},t) = T_1(\mathbf{x},t), \quad \tau_{ij}n_j = P_i, \quad i,j = 1,\ldots,N,,$$
$$\text{for } (t,\mathbf{x}) \in \Gamma_\tau(t) = I \times \cup_{\iota=1}^{r}(\Gamma_\tau \cap \partial\Omega^\iota), \tag{36}$$
$$\kappa_{ij}\frac{\partial T(\mathbf{x},t)}{\partial x_j}n_i = K(Y(\mathbf{x}) - T(\mathbf{x},t)) \quad \left(\text{or } \kappa_{ij}\frac{\partial T(\mathbf{x},t)}{\partial x_j}n_i = q\right),$$
$$u_i = {}^1u_{2i} \equiv 0,\ i = 1,\ldots,N, \quad \text{for } (t,\mathbf{x}) \in {}^1\Gamma_u(t) = I \times \cup_{\iota=1}^{r}({}^1\Gamma_u \cap \partial\Omega^\iota), \tag{37}$$

$$\kappa_{ij}\frac{\partial T(\mathbf{x},t)}{\partial x_j}n_i = q \ \ (\text{or} \equiv \mathrm{K}(\mathrm{Y}(\mathbf{x}) - \mathrm{T}(\mathbf{x},\mathrm{t}))),\ \mathrm{u_i} = {}^2\mathrm{u_{i2}} \neq 0,\ \mathrm{i,j} = 1,\ldots,\mathrm{N},$$
$$\text{on } {}^2\Gamma_u(t) = I \times \cup_{\iota=1}^{r}({}^2\Gamma_u \cap \partial\Omega^\iota), \text{where we denoted by } {}^2\Gamma_u \equiv \cup_{i=2}^{4}\,{}^i\Gamma_u, \tag{38}$$

and the contact conditions

$$\left.\begin{array}{ll}
(a) & T^s = T^m, \\
 & \kappa_{ij}\frac{\partial T}{\partial x_i}n_{j|(s)} - \kappa_{ij}\frac{\partial T}{\partial x_i}n_{j|(m)} = \mathcal{F}_c^{sm}|\tau_n^{sm}|\,|[\mathbf{u}_t']^{sm}|, \\
(b) & [u_n]^{sm} \le d^{sm},\ \tau_n^s = \tau_n^m \equiv \tau_n^{sm} \le 0, \\
 & ([u_n]^{sm} - d^{sm})\tau_n^{sm} = 0, \\
(c) & [\mathbf{u}_t']^{sm} = 0 \Longrightarrow |\boldsymbol{\tau}_t^{sm}| \le \mathcal{F}_c^{sm}(0)|\tau_n^{sm}|, \\
 & [\mathbf{u}_t']^{sm} \neq 0 \Longrightarrow \boldsymbol{\tau}_t^{sm} = -\mathcal{F}_c^{sm}(\mathbf{u}_t', T)|\tau_n^{sm}|\frac{[\mathbf{u}_t']^{sm}}{|[\mathbf{u}_t']^{sm}|}, \\
 & \text{and } \alpha = \alpha_0(\mathbf{x}) + \alpha_D(\mathbf{x},\lambda^s,D),
\end{array}\right\} (t,\mathbf{x}) \in I \times \cup_{s,m}\Gamma_c^{sm}, \tag{39}$$

$$T(\mathbf{x},0) = T_0(\mathbf{x}), \quad \mathbf{u}(\mathbf{x},0) = \mathbf{u}_0(\mathbf{x}), \quad \mathbf{u}'(\mathbf{x},0) = \mathbf{u}_1(\mathbf{x}), \quad \mathbf{x} \in \Omega, \tag{40}$$

where $\alpha = \alpha(\mathbf{x}) \ge 0$ represents the elastic wave damping in the investigated region and such that $\alpha = \alpha_0(\mathbf{x}) \ge 0$, that corresponds to the natural elastic wave damping in the investigated region, and moreover, such that at the moment when the frictional forces are not sufficient to preclude the mutual shifting of both geological bodies - the lithospheric plates and/or the lithospheric blocks in the subduction zone, then in the distances $\mathbf{x}$ from the focus (hypocenter) greater then $\mid X \mid$, i.e., near the boundary ${}^1\Gamma_u$, through the distance D the elastic wave damping will be $\alpha = \alpha_0(\mathbf{x}) + \alpha_D(\mathbf{x},\lambda^s,D)$, where α_D, that being a function of the wave length λ^s and the distance D from the boundary ${}^1\Gamma_u$, represents the elastic wave damping that fully damps all of the seismic waves propagated in the region near the boundary ${}^1\Gamma_u$, λ^s represents the (smallest) seismic wave length and D is the distance from the boundary ${}^1\Gamma_u$ (see Nedoma (1998)), and where $c_{ijkl}^{(e)*}$ and $c_{ijkl}^{(v)*}$ are the nonlinear elastic and viscous coefficients of the rocks defined by $c_{ijkl}^{(n)*} = 2\lambda\left[A^{(n)}(e_{ij}(\mathbf{u}))\right]^{\lambda-1} c_{ijkl}^{(n)}(\mathbf{x})$, $n = e,v$, where $c_{ijkl}^{(n)}$

satisfy [2]

$$c_{ijkl}^{(n)}(\mathbf{x}) \in L^{\infty}(\Omega), \quad c_{ijkl}^{(n)}(\mathbf{x}) = c_{jikl}^{(n)}(\mathbf{x}) = c_{klij}^{(n)}(\mathbf{x}) = c_{ijlk}^{(n)}(\mathbf{x}), \; n = e, v,$$
$$c_{ijkl}^{(n)}(\mathbf{x})e_{ij}e_{kl} \geq c_0^{(n)} e_{ij}e_{ij} \quad \forall e_{ij}, \quad e_{ij} = e_{ji} \text{ and a.e. } \mathbf{x} \in \Omega, \quad c_0^{(n)} > 0, \tag{41}$$
$$\text{for the anisotropic rock materials,}$$
$$c_{ijkl}^{(n)} = \lambda^{(n)}\delta_{ij}\delta_{kl} + \mu^{(n)}(\delta_{ik}\delta_{jl} + \delta_{il}\delta_{jk}) \text{ for the isotropic rock materials,}$$

κ_{ij} are the non-linear heat conduction depending on the space variables, the temperature and the gradient of temperature, that are continuous in the sense of Carathéodory (i.e., $\mathbf{x} \mapsto \kappa_{ij}(\mathbf{x}, T, \nabla T)$ is measurable for every $T \in \mathbb{R}$ and $\mathbf{x} \mapsto \kappa_{ij}(\mathbf{x}, T, \nabla T)$ is continuous for almost every $\mathbf{x} \in \Omega$) and satisfy the growth condition

$$\kappa_0(1 + \delta_0 \mid T \mid^{\gamma_0} + \delta_1 \mid \nabla T \mid^{\gamma_1})\xi_i\xi_j \leq \kappa_{ij}(\mathbf{x}, T, \nabla T) \leq$$
$$\leq \kappa_1(1 + \delta_0 \mid T \mid^{\gamma_0} + \delta_1 \mid \nabla T \mid^{\gamma_1})\xi_i\xi_j \tag{42}$$

for all $\{\xi_i\} \in \mathbb{R}$ and all $\mathbf{x} \in \Omega$ with $0 < \kappa_0 \leq \kappa_1 < +\infty$, $\gamma_0 \geq 1/2, \gamma_1 \geq 1/2, \delta_0 \geq 0, \delta_1 \geq 0$, and the strongly monotonicity condition and the continuity relation (Eck et al. (2005),

$q = q(\mathbf{x}, t)$ is the heat flux, K the heat transition coefficient, $Y(\mathbf{x})$ the external temperature, $T_0(\mathbf{x}), T(\mathbf{x}, t)$ the initial and actual temperatures, $\mathbf{u} = \mathbf{u}(\mathbf{x}, t)$, $\boldsymbol{\tau} = \boldsymbol{\tau}(\boldsymbol{x}, t) = \boldsymbol{\tau}(\boldsymbol{u}, T)$ are the displacement and the stress vectors, τ_{ij} the stress tensor, u_n, $\mathbf{u}_t$, $\tau_n, \boldsymbol{\tau}_t$ the normal and tangential components of displacement and stress vectors, $\mathbf{n}$ the outward normal to the boundary $\partial\Omega$ including Γ_c^{sm} and the term $\mathcal{F}_c^{sm}|\tau_n^{sm}|\,|[\mathbf{u}_t']^{sm}|$ represents the frictional heats on the contact surfaces Γ_c^{sm}, and where $\mathbf{u}_0$, $\mathbf{u}_1$, $\mathbf{u}_2$ are given functions, where by $\mathbf{u}_0$ and $\mathbf{u}_1$ we can simulate distributions of the stress-strain or the stress-strain rate fields that accumulated in the lithosphere from the last earthquake, respectively, and that describe the initial situation in the studied region Ω, and where by $\mathbf{u}_2 = ({}^1\mathbf{u}_1, {}^2\mathbf{u}_2)$ we simulate the conditions on the damped part of the boundary with respect to the above discussed elastic wave damping, i.e., ${}^1\mathbf{u}_2 = \mathbf{0}$, and the movements of the lithospheric plates (blocks) in the sense of the plate tectonic hypothesis or the motions of oceanic water and of the motion of water evoked by the typhoon, i.e., ${}^2\mathbf{u}_2 \neq \mathbf{0}$, and that $\mathbf{u}_2$ has a time derivative $\mathbf{u}_2'$. Generally, the coefficient of friction $\mathcal{F}_c^{sm} \equiv \mathcal{F}_c^{sm}(\mathbf{x}, \mathbf{u}')$ (with a compact support $\mathcal{SF}_c \equiv \text{supp}_{\mathbf{x}}(\mathcal{F}_c)$) is globally bounded, non-negative and satisfies the Carathéodory property, i.e., $\mathcal{F}_c^{sm}(., \mathbf{v})$ is measurable for all $\mathbf{v} \in \mathbb{R}$, and $\mathcal{F}_c^{sm}(\mathbf{x}, .)$ is continuous for a.e. $\mathbf{x} \in \Gamma_c^{sm}$ and it depends on the space variable $\mathbf{x}$ and since we model also the difference between the coefficients of friction and of stick as well as slip, it depends also on the tangential displacement rate component $\mathbf{u}_t'$ (Eck et al. (2005), Nedoma (2010), (2012)), and where $[\mathbf{u}]^{sm} = \mathbf{u}^s - \mathbf{u}^n$ is a jump (difference) of quantities $\mathbf{u}^s$ and $\mathbf{u}^m$ and d^{sm} is a gap, where

$$d^{sm}(\mathbf{x}) = \frac{\varphi^s(\mathbf{x}) - \varphi^m(\mathbf{x})}{\sqrt{1 + |\nabla\varphi^s(\mathbf{x})|^2}},$$

[2]In general case the elastic and viscous coefficients can be also a function of $\mathbf{u}$, (see Nedoma, Hlaváček (2002))

where $\varphi^s, \varphi^m \in C^1$, defined on an open subset Γ_c^{sm} of $\mathbb{R}^{N-1}$, are the functions parametrized the two contact boundaries, e.g. the contact of both lithospheric plates in the subduction zone in the first case, or the two opposite faces of the fractures in the lithosphere in the second case, respectively. Thus the terms $d^{sm} \geq 0$ are the normalized gaps between the contact boundaries of Ω^s and Ω^m (e.g. of both neighboring lithospheric plates or lithospheric blocks) and between the two faces of the fracture (fault) (i.e., Γ_c^s and Γ_c^m). Moreover, on $\cup_{s,m}\Gamma_c^{sm}$ due to the equilibrium of forces it holds $\tau_{ij}(\mathbf{u}^s, T^s)n_j^s = -\tau_{ij}(\mathbf{u}^m, T^m)n_j^m$. In the heat equation the second term on the right hand side represents the deformation heat and the third term represents the viscous heat. Furthermore, we denote by $(y)_+ \equiv \max\{y, 0\}$ the positive part of y. Next, we will assume that $\gamma_0 = \gamma_1 = \gamma$.

Remark 2. *Note, that in general the multibody problems with Coulomb friction in a continuous domain and formulated in displacements are up-to-date an open problem, only simpler problems were in detail analysed (Eck et al. (2005), Cocou and Scarella (2006) and some of others).*

The cases if $\lambda = 1, \delta_0 = 1, \delta_1 = 0$ *and if* $\lambda = 1, \delta_0 = 0, \delta_1 = 1$, $\gamma \geq 0.123$ *for* $N = 2$, $\gamma \geq \frac{1}{2}$ *for* $N = 3$*, are studied and discussed in the continuous domain in Eck et al. (2005).*

3.2. Variational Formulation of the Problem

Let us introduce the spaces $L^{p,N}(\Omega)$, $p \in [1, +\infty)$, $L^\infty(\Omega)$, the Sobolev spaces $H^{1,N}(\Omega)$, $H_0^{1,N}(\Omega)$, $H^{\frac{1}{2},N}(\Gamma_c)$, $H_{00}^{\frac{1}{2},N}(\Gamma_c)$ by the usual way (Adams (1975)), and let $B(M)$ be the space of bounded functions endowed with the sup norm, and let $H_p^{s,N}(\Omega)$, where $s > 0$ is a non-integer, $[s]$ its integer part, $1 \leq p \leq \infty$, $N = 2, 3$, be the Sobolev-Slobodetskii space, i.e., the linear space of all functions $f \in L^p(\mathbb{R}^N)$ with the norm

$$\| f \|_{H_p^{s,N}(\mathbb{R}^N)} = \| f \|_{L^{p,N}(\mathbb{R}^N)} + \sum_{|\alpha| \leq [s]} \left(\int_\Omega \int_\Omega \frac{| D^\alpha f(x) - D^\alpha f(y) |^p}{|x-y|^{(s-[s])p+N}} dxdy \right)^{1/p} < \infty,$$

and moreover, the spaces and sets

$$V_0 = \left\{ \mathbf{v} | \mathbf{v} \in \prod_{\iota=1}^{r} H^{1,N}(\Omega^\iota), \mathbf{v} = 0 \text{ a.e. on } \Gamma_u \right\},$$

$$V = \mathbf{u}_2 + V_0, \; \mathcal{V} = \mathbf{u}_2' + \mathcal{V}_0 = L^2(I; V), \; K = \{\mathbf{v} \in V | [v_n]^{sm} \leq d^{sm} \text{ a.e. on } \Gamma_c^{sm}\},$$

$$\mathcal{K} = \left\{ \mathbf{v} | \mathbf{v} \in L^2(I; \prod_{\iota=1}^{r} H^{1,N}(\Omega^\iota)), \mathbf{v} = \mathbf{u}_2' \text{ on } \Gamma_u(t), [v_n]^{sm} \leq d^{sm} \text{ a.e. on } \Gamma_c^{sm}(t) \right\},$$

$${}^1V_0 = \left\{ z | z \in \prod_{\iota=1}^{r} H_{2+\gamma}^{1,1}(\Omega^\iota), z = 0 \text{ a.e. on } \Gamma_\tau \right\}, \; {}^1V = T_1 + {}^1V_0, \; {}^1\mathcal{V} = L^{2+\gamma}(I; {}^1V),$$

where we put $\gamma = \gamma_1 = \gamma_2 \geq 1/2$.

Let $\rho^\iota \in C(\overline{\Omega}^\iota)$, $\rho^\iota \geq \rho_0^\iota > 0$, $c_{ijkl}^{(n)*} = 2\lambda \left[A^{(n)}(e_{ij}(\mathbf{u}))\right]^{\lambda-1} c_{ijkl}^{(n)}(\mathbf{x})$, $n = e, v$, where $c_{ijkl}^{(n)\iota}$ satisfy $c_{ijkl}^{(n)\iota} \in L^\infty(\Omega^\iota)$ and the above conditions, $\mathbf{F}^\iota, \mathbf{F}'^\iota \in L^2(I; L^{2,N}(\Omega^\iota))$, $\mathbf{P}, \mathbf{P}' \in L^2(I; L^{2,N}(\Gamma_\tau))$, $\mathbf{u}_0 \in K$, $\mathbf{u}_1 \in V$, $\mathbf{u}_2' \in L^2\left(I; \prod_{\iota=1}^r H^{1,N}(\Omega^\iota)\right)$, $d^{sm} \in H^{\frac{1}{2},N}(\Gamma_c^{sm})$, $d^{sm} \geq 0$ a.e. on Γ_c^{sm}, $\mathcal{F}_c^{sm} \in L^\infty(\Gamma_c^{sm})$, $\mathcal{F}_c^{sm} \geq 0$ a.e. on Γ_c^{sm}, $c_e^\iota \in C(\overline{\Omega}^\iota)$, $c_e^\iota \geq c_0^\iota > 0$, κ_{ij}^ι satisfy the above conditions, $\frac{\partial \beta_{ij}^\iota}{\partial x_j} \in L^\infty(\Omega^\iota)$ $\forall i, j \in \{1, \cdots, r\}$, $Y \geq 0, Y \in L^2(\Gamma_u)$, $K \geq 0, K \in L^\infty(\Gamma_u)$.

In some special cases then instead of the space $H^{\frac{1}{2},N}(\Gamma_c^{sm})$ we will use the space $H_{00}^{\frac{1}{2},N}(\Gamma_c^{sm})$, the interpolation space between the space $L^{2,N}(\Gamma_c^{sm})$ and the space $H_0^{1,N}(\Gamma_c^{sm})$.

Then multiplying (33) and (34) by $\mathbf{v} - \mathbf{u}'$ or by $z - T$, respectively, where $\mathbf{v}$ and z are suitable test functions, then integrating over $\Omega^\iota \times I$ for all ι, using the Green's theorems and boundary and contact conditions, then we obtain the following variational problem:

Problem $(\mathcal{P})_v$: Find a vector function $\mathbf{u} \in \mathcal{K}$ with $\mathbf{u}' \in \mathcal{K} \cap B(I; L^{2,N}(\Omega))$ and $\mathbf{u}(\cdot, 0) = \mathbf{u}_0$, $\mathbf{u}'(\cdot, 0) = \mathbf{u}_1$, and a scalar function $T \in {}^1\mathcal{V}$ and $T(., 0) = T_0$, such that for all $t \in I$

$$\begin{aligned}
&\int_I \Big\{ (\mathbf{u}''(t), \mathbf{v} - \mathbf{u}'(t)) + (\alpha \mathbf{u}'(t), \mathbf{v} - \mathbf{u}'(t)) + a^{(e)}(\mathbf{u}(t); \mathbf{u}(t), \mathbf{v} - \mathbf{u}'(t)) + \\
&+ a^{(v)}(\mathbf{u}'(t); \mathbf{u}'(t), \mathbf{v} - \mathbf{u}'(t)) + b_s(T(t) - T_0, \mathbf{v} - \mathbf{u}') + j(\mathbf{v}) - j(\mathbf{u}'(t)) \Big\} dt \geq \\
&\geq \int_I (\mathbf{f}(t), \mathbf{v} - \mathbf{u}'(t)) dt \quad \forall \mathbf{v} \in \mathcal{K}, \qquad (43) \\
&\int_I \Big\{ (T', z - T(t)) + a^{(T)}(T(t); T(t), z - T(t)) + b_p(\mathbf{u}(t), z - T(t)) + \\
&+ \langle K(T - Y), z - T(t) \rangle_{\Gamma_u} \} dt \geq \\
&\geq \int_I \Big\{ (W, z - T(t)) + \langle c_{ijkl}^{(v*)} e_{kl}(\mathbf{u}') e_{ij}(\mathbf{u}'), z - T(t) \rangle_\Omega + \\
&+ \langle \mathcal{F}_c^{sm} \mid \tau_n^{sm}(\mathbf{u}, T) \mid\mid \mathbf{u}_t' \mid, z - T(t) \rangle_{\Gamma_c^{sm}} \} dt \quad \forall z \in {}^1\mathcal{V}, \qquad (44)
\end{aligned}$$

hold, where

$$\begin{aligned}
&(\mathbf{u}'', \mathbf{v}) = \sum_{\iota=1}^r (\mathbf{u}''^\iota, \mathbf{v}^\iota) = \int_\Omega \rho u_i'' v_i d\mathbf{x}, \quad (\alpha \mathbf{u}', \mathbf{v}) = \sum_{\iota=1}^r (\alpha \mathbf{u}'^\iota, \mathbf{v}^\iota) = \int_\Omega \alpha u_i' v_i d\mathbf{x}, \\
&a^{(n)}(\mathbf{w}; \mathbf{u}, \mathbf{v}) = \sum_{\iota=1}^r a^{(n)\iota}(\mathbf{w}^\iota; \mathbf{u}^\iota, \mathbf{v}^\iota) = \int_\Omega c_{ijkl}^{(n)*}(\mathbf{x}, \mathbf{w}) e_{kl}(\mathbf{u}^\iota) e_{ij}(\mathbf{v}^\iota) d\mathbf{x}, \quad n = e, v, \\
&b_s(T(t) - T_0, \mathbf{v}) = \sum_{\iota=1}^r b_s^\iota(T^\iota(t) - T_0^\iota), \mathbf{v}^\iota) = \int_\Omega \frac{\partial}{\partial x_j}(\beta_{ij}(T(t) - T_0)) v_i d\mathbf{x}, \\
&(\mathbf{f}, \mathbf{v}) = \sum_{\iota=1}^r (\mathbf{f}^\iota, \mathbf{v}^\iota) = \int_\Omega \mathbf{F} \cdot \mathbf{v} d\mathbf{x} + \int_{\Gamma_\tau} \mathbf{P} \cdot \mathbf{v} ds,
\end{aligned}$$

$$j(\mathbf{v}) = \int_{\cup_{s,m}\Gamma_c^{sm}} \mathcal{F}_c^{sm} |\tau_n^{sm}(\mathbf{u},T)| ([\mathbf{v}_t]^{sm})\, ds,$$

$$(T',z) = \sum_{\iota=1}^{r} (T'^{\iota}, z^{\iota}) = \int_{\Omega} \rho c_e T' z\, d\mathbf{x},$$

$$a^{(T)}(w;T,z) = \sum_{\iota=1}^{r} a^{(T)\iota}(w^{\iota};T^{\iota},z^{\iota}) = \int_{\Omega} \kappa_{ij}(\mathbf{x},w,\nabla w) \frac{\partial T}{\partial x_i} \frac{\partial z}{\partial x_j} d\mathbf{x},$$

$$b_p(\mathbf{u},z) = \sum_{\iota=1}^{r} b^{\iota}(\mathbf{u}^{\iota},z^{\iota}) = \int_{\Omega} \rho \beta_{ij} T_0 \frac{\partial v_i}{\partial x_j} z\, d\mathbf{x},$$

$$(Q,z) = \sum_{\iota=1}^{r} (Q^{\iota},z^{\iota}) = \int_{\Omega} W z\, d\mathbf{x} + \int_{^2\Gamma_u} q z\, ds,$$

$$\langle K(T-Y),z\rangle_{^1\Gamma_u} = \sum_{\iota=1}^{r} \langle K^{\iota}(T^{\iota}-Y^{\iota}),z^{\iota}\rangle_{^1\Gamma_u \cap \partial\Omega^{\iota}} = \int_{^1\Gamma_u} K(T-Y) z\, ds,$$

$$\langle \mathcal{F}_c^{sm} \mid \tau_n^{sm}(\mathbf{u},T) \mid\mid [\mathbf{u}'_t]^{sm} \mid, z)\rangle_{\Gamma_c^{sm}} = \int_{\cup_{s,m}\Gamma_c^{sm}} \mathcal{F}_c^{sm} |\tau_n^{sm}(\mathbf{u},T)| \left|[\mathbf{u}'_t]^{sm}\right| z\, ds,$$

$$\langle c_{ijkl}^{(v)*} e_{kl}(\mathbf{u}') e_{ij}(\mathbf{u}'), z\rangle_{\Omega} = \sum_{\iota=1}^{r} \langle c_{ijkl}^{(v)*\iota} e_{kl}(\mathbf{u}'^{\iota}) e_{ij}(\mathbf{u}'^{\iota}), z\rangle_{\Omega} =$$

$$= \int_{\Omega} c_{ijkl}^{(v)*}(\mathbf{x},\mathbf{u}') e_{kl}(\mathbf{u}') e_{ij}(\mathbf{u}')\, z\, ds,$$

and where in the thermal part of the problem the growth condition, strong monotonicity condition and continuity relations are required. In Eck et al. (2005) and Cocou and Scarella (2006) dynamic contact problems in thermo-visco-elasticity with Coulomb friction and with heat conductivity depending on the temperature or on the gradient of temperature are analyzed and discussed, where for the elastic part of the problem the linear visco-elastic rheology is assumed and that the visco-elastic body is in contact with the foundation that is assumed to be absolutely rigid. Since the friction is a source of heat so that it cannot be neglected. Moreover, because in the plate tectonic problems colliding lithospheric plates and lithospheric blocks in subduction zones have large contact areas, and because the process is relatively fast, then the thermal effects are very significant. This is the reason why we need to study the coupling of the contact problem with thermal problem, characterized by coupling of the equation of motion and the heat equation (Nedoma (1998)).

3.3. Numerical Solution of the Problem

In the numerical analyzes we simplified the problem by such a way that firstly we linearized the non-linear contact problem by using e.g., the secant modulus method or by the contraction method or by the steepest descent method and secondly then we introduce an iterative algorithm for approximation of the contact problems with Coulomb friction by the contact problem with the given friction (known as the Tresca model of friction or dry friction model).

Then as a useful algorithm the fixpoint algorithm for the Tresca friction will be used, where the Coulomb law of friction in every time level will be approximated by its value $g_c^{sm} \equiv \mathcal{F}_c^{sm}|\tau_n^{sm}(\mathbf{u},T)|(t-\Delta t)$ from the previous time level. Such algorithms are convergent for small coefficient of friction (Nečas et al. (1980), Eck et al. (2005)). The other algorithm can be based on the full Newton method (Hüeber et al. (2008)).

Next, for linearization of the non-linear problem we will introduce **the secant-modulus (Kachanov) method for variational inequalities**.

Firstly, let us introduce the following assumptions:

Presumptions:

(i) Let K be a nonempty closed convex set in the real Hilbert space V, let $b \in V'$ and let $u_0 \in K$ be given.

(ii) Let the functional $L : K \to \mathbb{R}$ be G-differentiable. Then there exists a functional $B : K \times V \times V \to \mathbb{R}$ such that, for $u, v, w \in K$, we have the representation formula

$$\langle L'(u), v - w \rangle = B(u; u, v - w) \tag{45}$$

and the key inequality

$$L(v) - L(u) \leq 2^{-1}(B(u; v, v) - B(u; u, u)) . \tag{46}$$

(iii) For each $u \in K$, the map $(v, w) \mapsto B(u; v, w)$ is bilinear, bounded, and symmetric from $V \times V$ to $\mathbb{R}$. There exist numbers $\rho > 0$ and $\delta > 0$ such that

$$|B(u; v, w)| \leq \delta \|v\| \|w\| \quad \text{for all } u \in K, v, w \in V$$

and

$$B(u; w - v, w - v) \geq \rho \|w - v\|^2 \quad \text{for all } u, v, w \in K .$$

(iv) The operator $L' : K \to V'$ is continuous and strongly monotone, i.e., there is a number $\rho_0 > 0$ such that

$$\langle L'(u) - L'(v), u - v \rangle \geq \rho_0 \|u - v\|^2 \quad \text{for all } u, v \in K .$$

Theorem 1. *Let (i)-(iii) hold. Let* $j : V \to \overline{\mathbb{R}}$ *be weakly lower semicontinuous,*

$$j(v) \geq 0 \quad \forall v \in V \quad \textit{and} \quad |j(u) - j(v)| \leq C \|u - v\| \quad \forall u, v \in V .$$

Then:

(a) for $k = 0, 1, \ldots$*, the quadratic variational problem*

$$2^{-1} B(u_k; u_{k+1}, u_{k+1}) - b(u_{k+1}) = \min! \quad u_{k+1} \in K , \tag{47}$$

has a unique solution u_{k+1}*, and*

$$B(u_k; u_{k+1}, v - u_{k+1}) - j(v) - j(u_{k+1}) \geq b(v - u_{k+1}) \quad \textit{for all } v \in K . \tag{48}$$

(b) If, in addition, (iv) holds, then the original variational problem

$$L(u) - b(u) = \min! \quad u \in K,$$

has a unique solution u. Moreover, u is the unique solution of the variational inequality

$$\langle L'(u), v-u \rangle + j(v) - j(u) \geq b(v-u) \quad \textit{for all } v \in K \textit{ and fixed } u \in K, \tag{49}$$

and the secant modules method converges, that is

$$u_k \longrightarrow u \quad \textit{in } V \quad \textit{as } k \longrightarrow \infty.$$

For the proof see Han, Jensen, Shimansky (1997); for $j \equiv 0$ see Zeidler (1990).

Corollary 1. *Let (i)-(iii) hold. Suppose that the set K is compact and that the map*

$$u \mapsto B(u; v, w)$$

is equicontinuous on K with respect to all $v, w \in V$.

Then the sequence (u_k) has at least one cluster point and each such cluster point u is a solution of the variational inequality (48) (with $j \equiv 0$, see Zeidler (1990)).

Remark 3. *(a) We say that the sequence $\{f_n\}$ is* **equicontinuous** *at the point x iff (if and only if) for each $\varepsilon > 0$, there exists a $\delta(\varepsilon) > 0$ such that*

$$d(x,y) < \delta(\varepsilon) \quad \textit{implies} \quad \|f_n(x) - f_n(y)\| < \varepsilon \quad \textit{for all } n,$$

δ does not depend on n. We call (f_n) equicontinuous on M iff (f_n) is equicontinuous at each point $x \in M$ (Zeidler (1990), p.189).
(b) A point $u \in X$ is a **cluster point** *of a sequence $\{u_k\}$, $k \in \mathbb{N}$, if for every neighborhood X_0 of u, there are infinitely many natural numbers k such that $u_k \in X_0$ (i.e., u is a limit of some subsequence of the sequence $\{u_k\}$, $k \in \mathbb{N}$).*

Applying the secant modulus method then the problem investigated leads to the solution of a sequence of the coupled system of hyperbolic variational inequalities of semi-coercive type with variable coefficients and parabolic variational equations of the following form:

Problem $(\mathcal{P}_0)_{(SM)v}$: Find a pair of functions $(\mathbf{u}_m, T_m)$, $\mathbf{u}_m \in K, T_m \in {}^1V, m = 1, 2, \cdots$, such that

$$\begin{aligned}
&(\mathbf{u}''_{m+1}(t), \mathbf{v} - \mathbf{u}'_{m+1}(t)) + (\alpha \mathbf{u}'_{m+1}(t), \mathbf{v} - \mathbf{u}'_{m+1}(t)) + \\
&+ a^{(e)}(\mathbf{u}_m; \mathbf{u}_{m+1}(t), \mathbf{v} - \mathbf{u}'_{m+1}(t)) + a^{(v)}(\mathbf{u}'_m; \mathbf{u}'_{m+1}(t), \mathbf{v} - \mathbf{u}'_{m+1}(t)) + \\
&+ b_s(T_{m+1} - T_0, \mathbf{v} - \mathbf{u}'_{m+1}(t)) + j(\mathbf{v}) - j(\mathbf{u}'_{m+1}(t)) \geq \\
&\geq (\mathbf{f}(t), \mathbf{v} - \mathbf{u}'_{m+1}(t)) \quad \forall \mathbf{v} \in K,
\end{aligned} \tag{50}$$

$$\begin{aligned}
&(T'_{m+1}, z - T_{m+1}(t)) + a^{(T)}(T_m; T_{m+1}(t), z - T_{m+1}(t)) + b_p(\mathbf{u}_{m+1}, z - T_{m+1}(t)) + \\
&+ \langle K(T_{m+1} - Y), z - T_{m+1}(t) \rangle_{{}^1\Gamma_u^{sm}} \geq \\
&\geq (W, z - T_{m+1}(t)) + \langle c_{ijkl}^{(v)*}(\mathbf{u}'_m) e_{kl}(\mathbf{u}'_m) e_{ij}(\mathbf{u}'_m), z - T_{m+1}(t) \rangle_\Omega + \\
&+ \langle \mathcal{F}_c^{sm} \mid \tau_n^{sm}(\mathbf{u}_m, T_m) \mid\mid \mathbf{u}'_{t,m} \mid, z - T_{m+1}(t) \rangle_{\Gamma_c^{sm}} \quad \forall z \in {}^1V,
\end{aligned} \tag{51}$$

with corresponding initial conditions hold, where

$$(\mathbf{u}''_m, \mathbf{v}) = \int_\Omega \rho u''_{i,m} v_i d\mathbf{x}, \quad (\alpha \mathbf{u}'_m, \mathbf{v}) = \int_\Omega \alpha u'_{i,m} v_i d\mathbf{x},$$

$$a^{(n)}(\mathbf{u}_{m-1}; \mathbf{u}_m, \mathbf{v}) = \int_\Omega c^{(n)}_{ijkl}(\mathbf{u}_{m-1}) e_{kl}(\mathbf{u}_m) e_{ij}(\mathbf{v}) d\mathbf{x}, \quad n = e, v,$$

$$b_s(T_m - T_0, \mathbf{v}) = \int_\Omega \frac{\partial}{\partial x_j}(\beta_{ij}(T_m - T_0)) v_i d\mathbf{x},$$

$$(\mathbf{f}, \mathbf{v}) = \int_\Omega \mathbf{F} \cdot \mathbf{v} d\mathbf{x} + \int_{\Gamma_\tau} \mathbf{P} \cdot \mathbf{v} ds,$$

$$j(\mathbf{v}) = \int_{\cup_{s,m}\Gamma_c^{sm}} \mathcal{F}_c^{sm} |\tau_n^{sm}(\mathbf{u}_m, T_m)| (|[\mathbf{v}_t]^{sm}|) ds,$$

$$(T'_m, z) = \int_\Omega \rho c_e T'_m z d\mathbf{x},$$

$$a^{(T)}(T_{m-1}; T_m, z) = \int_\Omega \kappa_{ij}(\mathbf{x}, T_{m-1}, \nabla T_{m-1}) \frac{\partial T_m}{\partial x_i} \frac{\partial z}{\partial x_j} d\mathbf{x},$$

$$b_p(\mathbf{u}_m, z) = \int_\Omega \rho \beta_{ij} T_0 \frac{\partial u_{i,m}}{\partial x_j} z d\mathbf{x},$$

$$(Q, z) = \int_\Omega W z d\mathbf{x} + \int_{{}^1\Gamma_u} q z ds,$$

$$\langle K(T_m - Y), z \rangle_{{}^1\Gamma_u} = \int_{{}^1\Gamma_u} K(T_m - Y) z ds,$$

$$\langle \mathcal{F}_c^{sm} \mid \tau_n^{sm}(\mathbf{u}_{m-1}, T_{m-1}) \mid\mid [\mathbf{u}'_{t,m-1}]^{sm} \mid, z) \rangle_{\Gamma_c^{sm}} =$$

$$= \int_{\cup_{s,m}\Gamma_c^{sm}} \mathcal{F}_c^{sm} \mid \tau_n^{sm}(\mathbf{u}_{m-1}, T_{m-1}) \mid\mid [\mathbf{u}'_{t,m-1}]^{sm} \mid z ds,$$

$$\langle c^{(v)*}_{ijkl}(\mathbf{x}, \mathbf{u}'_{m-1}) e_{kl}(\mathbf{u}'_{m-1}) e_{ij}(\mathbf{u}'_{m-1}), z \rangle_\Omega =$$

$$= \int_\Omega c^{(v)*}_{ijkl}(\mathbf{x}, \mathbf{u}'_{m-1}) e_{kl}(\mathbf{u}'_{m-1}) e_{ij}(\mathbf{u}'_{m-1}) z ds$$

3.4. Approximation of the Problem by the Tresca Model of Friction

Let us assume that in the linearized problem the Coulomb law of friction in every time level is approximated by its value g_c^{sm} from the previous time level, that is, $g_c^{sm} \equiv \mathcal{F}_c^{sm}(\mathbf{u}'_{t,m})|\tau_n^{sm}(\mathbf{u}_m, T_m)|)(t - \Delta t)$. Thus g_c^{sm} is a non-negative function and has a meaning of a given friction limit (or a given friction bound, representing the magnitude of the limiting friction traction at which slip originates), and where $-g_c^{sm}$ has a meaning of a given frictional force, and Δt is a time element. Thus this problem is approximated by another problem in which in every time level we will solve the dynamic contact problem with the given friction, known as the Tresca model of friction. Denoting $(\mathbf{U}_m, \overline{T}_m)$ the corresponding solution, then we have to solve the following Tresca problem:

Problem $(\mathcal{P})_{(T)v}$**:** Find a pair of functions $(\mathbf{U}_m, \overline{T}_m)$, $\mathbf{U}_m \in K$, $\overline{T}_m \in {}^1V$, $m = 1, 2, \cdots$, such that

$$\begin{aligned}
&(\mathbf{U}''_{m+1}(t), \mathbf{v} - \mathbf{U}'_{m+1}(t)) + (\alpha \mathbf{U}'_{m+1}(t), \mathbf{v} - \mathbf{U}'_{m+1}(t)) + \\
&+ a^{(e)}(\mathbf{U}_m; \mathbf{U}_{m+1}(t), \mathbf{v} - \mathbf{U}'_{m+1}(t)) + a^{(v)}(\mathbf{U}'_m; \mathbf{U}'_{m+1}(t), \mathbf{v} - \mathbf{U}'_{m+1}(t)) + \\
&+ b_s(\overline{T}_{m+1} - T_0, \mathbf{v} - \mathbf{U}'_{m+1}) + j(\mathbf{v}) - j(\mathbf{U}'_{m+1}(t)) \geq \\
&\geq (\mathbf{f}(t), \mathbf{v} - \mathbf{U}'_{m+1}(t)) \quad \forall \mathbf{v} \in K, \qquad (52)\\
&(\overline{T}'_{m+1}, z - \overline{T}_{m+1}(t)) + a^{(T)}(\overline{T}_m; \overline{T}_{m+1}(t), z - \overline{T}_{m+1}(t)) \\
&+ b_p(\mathbf{U}_{m+1}, z - \overline{T}_{m+1}(t)) + \\
&+ \langle K(\overline{T}_{m+1} - Y), z - \overline{T}_{m+1}(t)\rangle_{\Gamma_u^{sm}} \geq (W, z - \overline{T}_{m+1}(t)) + \\
&+ \langle c^{(v)*}_{ijkl}(\mathbf{U}'_m) e_{kl}(\mathbf{U}'_m) e_{ij}(\mathbf{U}'_m), z - \overline{T}_{m+1}(t)\rangle_\Omega + \\
&+ \langle g_c^{sm} \mid [\mathbf{U}'_{t,m}]^{sm} \mid, z - \overline{T}_{m+1}(t)\rangle_{\Gamma_c^{sm}} \quad \forall z \in {}^1V, \qquad (53)
\end{aligned}$$

hold, where

$$\begin{aligned}
&(\mathbf{U}''_m, \mathbf{v}) = \int_\Omega \rho U''_{i,m} v_i d\mathbf{x}, \quad (\alpha \mathbf{U}'_m, \mathbf{v}) = \int_\Omega \alpha U'_{i,m} v_i d\mathbf{x}, \\
&a^{(n)}(\mathbf{U}_{m-1}; \mathbf{U}_m, \mathbf{v}) = \int_\Omega c^{(n)}_{ijkl}(\mathbf{U}_{m-1}) e_{kl}(\mathbf{U}_m) e_{ij}(\mathbf{v}) d\mathbf{x}, \quad n = e, v, \\
&b_s(\overline{T}_m - T_0, \mathbf{v}) = \int_\Omega \frac{\partial}{\partial x_j}(\beta_{ij}(\overline{T}_m - T_0)) v_i d\mathbf{x}, \\
&(\mathbf{f}, \mathbf{v}) = \int_\Omega \mathbf{F} \cdot \mathbf{v} d\mathbf{x} + \int_{\Gamma_\tau} \mathbf{P} \cdot \mathbf{v} ds, \quad j(\mathbf{v}) = \int_{\bigcup_{s,m} \Gamma_c^{sm}} g_c^{sm} |[\mathbf{v}_t]^{sm}| ds,
\end{aligned}$$

$$\begin{aligned}
&(\overline{T}'_m, z) = \int_\Omega \rho c_e \overline{T}'_m z d\mathbf{x}, \\
&a^{(T)}(\overline{T}_{m-1}; \overline{T}_m, z) = \int_\Omega \kappa_{ij}(\overline{T}_{m-1}, \nabla \overline{T}_{m-1}) \frac{\partial \overline{T}_m}{\partial x_i} \frac{\partial z}{\partial x_j} d\mathbf{x}, \\
&b_p(\mathbf{U}_m, z) = \int_\Omega \rho \beta_{ij} T_0 \frac{\partial U_{i,m}}{\partial x_j} z d\mathbf{x}, \quad (Q, z) = \int_\Omega W z d\mathbf{x} + \int_{{}^1\Gamma_u} q z ds, \\
&\langle K(\overline{T}_m - Y), z\rangle_{\Gamma_u} = \int_{\Gamma_u} K(\overline{T}_m - Y) z ds, \\
&\langle g_c^{sm} \mid [\mathbf{U}'_{t,m-1}]^{sm} \mid, z)\rangle_{\Gamma_c^{sm}} = \int_{\bigcup_{s,m} \Gamma_c^{sm}} g_c^{sm} \mid [\mathbf{U}'_{t,m-1}]^{sm} \mid z ds, \\
&\langle c^{(v)*}_{ijkl}(\mathbf{U}'_{m-1}) e_{kl}(\mathbf{U}'_{m-1}) e_{ij}(\mathbf{U}'_{m-1}), z\rangle_\Omega = \\
&= \int_\Omega c^{(v)*}_{ijkl}(\mathbf{U}'_{m-1}) e_{kl}(\mathbf{U}'_{m-1}) e_{ij}(\mathbf{U}'_{m-1}) z ds,
\end{aligned}$$

with $\mathbf{U}_m(\mathbf{x}, 0) = \mathbf{u}_0(\mathbf{x})$, $\mathbf{U}'_m(\mathbf{x}, 0) = \mathbf{u}_1(\mathbf{x})$ and $\overline{T}_m(\mathbf{x}, 0) = T_0(\mathbf{x})$, and where the bilinear forms $a^{(n)}(\mathbf{w}, \mathbf{u}, \mathbf{v})$, $n = e, v$, and $a^T(w, T, z)$ satisfy the same conditions as above.

To prove the existence of the solution of Problem $(\mathcal{P})_{(T)v}$ the useful decomposition, the penalty and regularization techniques can be used (Eck et al. (2005)).

3.4.1. Algorithm

We saw above that the problem leads to solve a coupled system of the hyperbolic variational inequality and the parabolic variational equation. Then the aim is to transform the non-linear problem into a sequence of linear problems. For the numerical solution the secant modulus method and the Tresca model of friction were used, that is, we approximated the frictional term by a given friction limit (Tresca model of friction). The algorithm will be based on the semi-implicit scheme in time and the finite elements in space and the PDAS method. For simplicity we can assume that the coefficient of friction is a function of the displacement only.

Let a region Ω be approximated by Ω_h, being a polyhedron for the 3D case, with the boundary $\partial\Omega_h = \Gamma_{\tau h} \cup \Gamma_{uh} \cup \Gamma_{ch}$. Let $\{\mathcal{T}_h\}$ be a regular family of finite element partitions $\mathcal{T}_h$ of $\overline{\Omega}_h$ compatible to the boundary subsets $\overline{\Gamma}_{\tau h}$, $\overline{\Gamma}_{uh}$ and $\overline{\Gamma}_{ch}$. Let $V_h \subset V$ be the finite element space of linear elements corresponding to the partition $\mathcal{T}_h$. Then $K_h = V_h \cap K$ is the set of continuous piecewise linear functions that vanish at the nodes of $\overline{\Gamma}_{uh}$ and whose normal components are non-positive at the nodes on $\cup_{s,m}\Gamma_c^{sm}$. It is evident that K_h is a nonempty, closed, convex subset of $V_h \subset V$. Let $\mathbf{u}_{0h} \in K_h$, $\mathbf{u}_{1h} \in K_h$ be an approximation of $\mathbf{u}_0$ and $\mathbf{u}_1$. Further, we assume that the end points $\overline{\Gamma}_{\tau h} \cup \overline{\Gamma}_{uh}$, $\overline{\Gamma}_{uh} \cup \overline{\Gamma}_{ch}$, $\overline{\Gamma}_{\tau h} \cup \overline{\Gamma}_{ch}$ coincide with the vertices of T_{hi}. Moreover, we will assume that the frictional term is approximated by its value in the previous time level(s), so that the frictional term at every $t \in I$ is approximated by a given friction limit. By a similar way we introduce the space ${}^1V_h \subset {}^1V$ as the finite element space of linear elements corresponding to the partition $\mathcal{T}_h$. Let $T_{0h} \in {}^1V_h$ be an approximation of T_0.

Then we have the following semi-discrete problem:

Let $I = (0,t_p), t_p > 0$, let $M > 0$ be an integer, then $\Delta t = t_p/M$, $t_i = i\Delta t$, $i = 0,\ldots,M$. Approximating the time derivatives by the differences, i.e., $\mathbf{U}''_{m,h} = \frac{\mathbf{U}^{i+1}_{m,h} - 2\mathbf{U}^i_{m,h} + \mathbf{U}^{i-1}_{m,h}}{\Delta t^2}$, $\mathbf{U}'_{m,h} = \frac{\mathbf{U}^{i+1}_{m,h} - \mathbf{U}^{i-1}_{m,h}}{2\Delta t}$, and where we set $\mathbf{U}^i_{m,h} = \mathbf{U}_{m,h}(t_i)$, $\Delta\mathbf{U}^i_{m,h} = \mathbf{U}_{m,h}(t_i) - \mathbf{U}_{m,h}(t_{i-1})$, $g^{sm}_{ch} = g^{sm}_{ch}(t_i) = \mathcal{F}^{sm}_c\left(\mathbf{U}^i_{m,h}, \overline{T}^i_{m,h}\right)\left|\tau^{sm}_n\left(\mathbf{U}^i_{m,h}, \overline{T}^i_{m,h}\right)\right|$, $(\mathbf{F}(t_{i+1}), \mathbf{v}_h) = \Delta t^2(\mathbf{f}_h(t_{i+1}), \mathbf{v}_h) + (2\mathbf{U}^i_{m,h} - \mathbf{U}^{i-1}_{m,h}, \mathbf{v}_h) + \frac{1}{2}\Delta t(\alpha\mathbf{U}^{i-1}_{m,h}, \mathbf{v}_h) + \Delta t^2 a^{(e)}_h(\mathbf{U}^{i-1}_{m-1,h}; \mathbf{U}^i_{m,h}, \mathbf{v}_h) + \Delta t a^{(v)}_h(\Delta\mathbf{U}^{i-1}_{m-1,h}; (\Delta t)^{-1}\mathbf{U}^{i-1}_{m,h}, \mathbf{v}_h) + \Delta t b_s(\overline{T}^i_{m,h} - T_0, \mathbf{v}_h) + \Delta t\langle g^{sm}_c, | \ [\mathbf{v}_t]^{sm} \ |\rangle_{\Gamma^{sm}_c}$, $\mathbf{F}(t_{i+1}) \equiv \mathbf{f}_h$, and similarly $\overline{T}'_{m,h} = \frac{\overline{T}^{i+1}_{m,h} - \overline{T}^i_{m,h}}{\Delta t}$, and then setting $\overline{T}^i_{m,h} = \overline{T}_{m,h}(t_i)$, $(Q(t_{i+1}), z_h) = \Delta t(W(t_{i+1}), z_h) + (\overline{T}^i_{m,h}, z_h) + \Delta t a^{(T)}(\overline{T}^i_{m-1,h}; \overline{T}^i_{m,h}, z_h) + \Delta t b_p(\mathbf{U}^i_{m,h}, z_h) + \Delta t\langle c^{(v)*}_{ijkl}(\mathbf{x}, \Delta\mathbf{U}^i_{m-1}/\Delta t)e_{kl}(\Delta\mathbf{U}^i_{m-1,h}/\Delta t)e_{ij}(\Delta\mathbf{U}^i_{m-1,h}/\Delta t), z_h\rangle_\Omega + \Delta t\langle g^{sm}_c \mid \mathbf{U}^i_{t,m} \mid, z\rangle_{\Gamma^{sm}_c}$ and putting $\mathbf{U}^{i+1}_{m,h} \equiv \mathbf{U}_h, \overline{T}^{i+1}_{m,h} \equiv \overline{T}_h$, $j(\mathbf{v}_h) = \Delta t^2 \int_{\cup_{s,m}\Gamma^{sm}_c} g^{sm}_{ch} |[\mathbf{v}_{ht}]^{sm}| ds$, where g^{sm}_{ch} is the approximate frictional limit.

Remark 4. *In the case if the coefficient of fricton depends on velocities, then* $g^{sm}_{ch} = g^{sm}_{ch}(t_i) = \mathcal{F}^{sm}_c\left(\Delta t^{-1}[\Delta\mathbf{U}^i_{m,h}]^{sm}, [\overline{T}^i_{m,h}]^{sm}\right)\left|\tau^{sm}_n\left(\mathbf{U}^i_{m,h}, \overline{T}^i_{m,h}\right)\right|$.

Next, for simplicity we firstly set $\overline{\mathbf{U}}_h = \mathbf{U}^{i-1}_{m-1,h}$ and $\overline{\overline{T}}_h = \overline{T}^{i-1}_{m-1,h}$, and then we will omit the index m from the secant modulus method and the index $i+1$ from the time discrete approximation. Then after some modification we have to solve in every time level $t = t_{i+1}$ the following problem:

Problem $(\mathcal{P}_A)_h$**:** Find $\mathbf{U}_h \in K_h$, $\overline{T}_h \in {}^1V_h$, a.e. $t = t_{i+1} \in I$, such that

$$A(\mathbf{U}_h, \mathbf{v}_h - \mathbf{U}_h) + j(\mathbf{v}_h) - j(\mathbf{U}_h) \geq (\mathbf{f}_h, \mathbf{v}_h - \mathbf{U}_h), \qquad \forall \mathbf{v}_h \in K_h,\ t = t_{i+1} \in I, \tag{54}$$

$$A^{(T)}(\overline{T}_h, z_h - \overline{T}_h) \geq (Q_h, z_h - T_h), \qquad \forall \overline{T}_h \in {}^1V_h,\ t = t_{i+1} \in I, \tag{55}$$

where

$$A(\mathbf{U}_h, \mathbf{v}_h) = (\mathbf{U}_h, \mathbf{v}_h) + \Delta t^2 a_h^{(e)}(\overline{\mathbf{U}}_h; \mathbf{U}_h, \mathbf{v}_h) + \Delta t a_h^{(v)}(\overline{\mathbf{U}}_h; \mathbf{U}_h, \mathbf{v}_h)$$

for the visco-elastic part and

$A^{(T)}(\overline{T}_h, z_h) = (\overline{T}_h, z_h) + \Delta t a_h^{(T)}(\overline{\overline{T}}_h; \overline{T}_h, z_h)$ for the thermal part.

According to the above assumptions about the bilinear forms $a_h^{(n)}(\cdot;\cdot,\cdot), n = e, v, a_h^{(T)}(\cdot;\cdot,\cdot)$ and since $\rho \geq \rho_0 > 0,\ c_e \geq c_{e0} > 0$, then the bilinear forms $A(\cdot,\cdot)$ and $A^T(\cdot,\cdot)$ satisfy

$$\begin{aligned}
A(\mathbf{v}_h, \mathbf{v}_h) &\geq A_0 \|\mathbf{v}_h\|_{1,N}^2, \quad A_0 = \text{const.} > 0, \\
|A(\mathbf{v}_h, \mathbf{w}_h)| &\leq A_1 \|\mathbf{v}_h\|_{1,N} \|\mathbf{w}_h\|_{1,N}, \quad A_1 = \text{const.} > 0,\ \mathbf{v}_h, \mathbf{w}_h \in V_h, \\
A^T(z_h, z_h) &\geq A_0^T \|z_h\|_{1,N}^2, \quad A_0^T = \text{const.} > 0, \\
|A(z_h, w_h)| &\leq A_1^T \|z_h\|_{1,1} \|w_h\|_{1,1}, \quad A_1^T = \text{const.} > 0,\ z_h, w_h \in {}^1V_h.
\end{aligned}$$

To prove the convergence of this algorithm, we need to prove the stability and the consistency of the used schemes. Most of the stability theory of the numerical methods for nonlinear problems is concerned with linearized problems. It was shown that if a numerical algorithm is unstable for the linear problem then it is unstable also for the nonlinear problem because linear problems are subsets of nonlinear problems.

By the usual ways the finite element approximations lead to solve the coupled system of the discrete equation of motion for a damped system and the discrete equation of heat conduction. Hence, by the usual ways the displacements and the temperatures measured in a local coordinate system x, y, z within each element T_{hi} will be found and then we will determine the displacements and temperatures at the n finite element nodal points. In the visco-elastic case the displacement vector includes the displacements and the rotations, therefore, we need to determine the values of the virtual displacements. The total internal virtual work is equal to the total external virtual work over all finite elements, then under the assumption that the element accelerations and element velocities are approximated in the same way as the element displacements, and similarly in the temperature case, we obtain the coupled system of the ordinary differential equations of the second and the first orders of the form

$$M\mathbf{U}_h'' + C\mathbf{U}_h' + K\mathbf{U}_h = \mathbf{F}_h, \tag{56}$$

$$C^{(T)}\mathbf{T}_h' + K^{(T)}\mathbf{T}_h = \mathbf{Q}_h, \tag{57}$$

where M is the acceleration matrix, C is the damping matrix, K is the stiffness matrix, $\mathbf{U}_h$ is the vector of virtual displacements, $\mathbf{F}_h$ are the volume, surface, frictional, thermal, etc., forces, $C^{(T)}$ is the capacitance matrix, $K^{(T)}$ is the conductance matrix and $\mathbf{Q}_h$ is the thermal source vector and $\mathbf{T}_h$ is the vector of nodal temperatures. Because the maximum eigenvalue (the spectral radius of the matrix) is the most restrictive on the size of the time step, then it yields the critical time step.

The discretization error will be a function of the time step Δt and the mesh size h and thus the truncation error of the time and spatial discretization must tend to zero (Bathe (1996), Belytschko et al. (2000), Nedoma (2010), (2012). Nedoma et al. (2011)). From the stability analysis for the critical time step size we have

$$\Delta t \leq \Delta t_{\text{crit}} \simeq \gamma \frac{h^{(n)}}{\pi} \left(\frac{\rho^{(n)}}{E^{(n)}} \right) [(\xi_k^2 - 1)^{1/2} - \xi_k] \tag{58}$$

for the non-linear elastic part and

$$\Delta t \leq \Delta t_{\text{crit}}^{(T)} \simeq \gamma \frac{2}{(1-2\overline{\alpha})\lambda_{max_k}} \quad \text{for the thermal part}, \tag{59}$$

where $h^{(n)}$ is the diameter of the corresponding (n)-th element, $h^{(n)} = c^{(n)} T_n$, T_n is the smallest period of the finite discretization with n degrees of freedom, $c^{(n)}$ is a dilatational wave velocity in the (n)-th material element, $\rho^{(n)}$ and $E^{(n)}$ are (average) values of the density and the Young modulus on the (n)-th element, and γ is a reduction factor determining from numerical experiments, and the damping is usually given as a fraction of critical damping over a k-th element $\xi_k = \frac{a_1}{2\omega_k} + \frac{a_2\omega_k}{2}$, and where the factor in the bracket in (58) is equal to 1 for $\xi_k = 0$ and less than 1 for $\xi_k > 0$. Hence, the delay in the velocity decreases the stable time step when the damping is assumed. The critical time step is also called the stable time step. Moreover, the algorithm is also consistent of order two, because the truncation error is of order Δt^2 in the displacements. Furthermore, $\lambda_{max_k} = \max_k \lambda_k \leq 1$, λ_k are the eigenvalues of the matrix $K^{(T)}$, $0 \leq \overline{\alpha} < \frac{1}{2}$. For $\overline{\alpha} = 0$ considering the explicit form of the generalized trapezoidal rule we have the forward Euler method. As the result we see that the algorithm is convergent.

The system (57) is a system of ordinary differential equations of the first order and it can be also solved by the mortar method(s) presented in Falletta and Lamichhane (2009) or methods presented in McDevitt and Laursen (2000), Miehe and Wriggers (1994), Wriggers (2002) and Laursen (2002)).

In the next step the discussed algorithm will be based on (i) a saddle point formulation, the mortar discretization and the primal-dual active set technique for the case with the Coulomb friction in the elastic part of the problem and (ii) on the mortar method for the thermal part of the problem. Because our main interests are the physical problems connected with the tsunamiearthquake and propagation of the seismic waves and with the ensuing tsunami waves, both under the typhoon, and then their consequences, we limit ourselves to these problems only. The thermal part of the solved problem is substantial in principle in the case of an effect of volcanism or in the case of the global plate tectonic problems, and therefore, we will discuss the problem only shortly and we refer to the above cited papers and books.

The mortar discretization of the equation of motion

To give a saddle point formulation it is usually to introduce a Lagrange multiplier space $M = M_n \times M_t$, being the dual space of the trace space $W = \prod_s H^{\frac{1}{2},N}(\Gamma_c^s)$ (i.e., the trace space of V_0 restricted to $\cup_s \Gamma_c^s$) and its dual is $W' = \prod_s H^{-\frac{1}{2},N}(\Gamma_c^s)$, under the assumption that Ω^ι, $\iota = 1,\ldots,r$, are domains with sufficiently smooth boundaries $\partial\Omega^\iota$. In the special cases we must use the space $H_{00}^{\frac{1}{2},N}(\Gamma_c^s)$, i.e., the interpolation space between $L^{2,N}(\Gamma_c^{sm})$ and $H_0^{1,N}(\Gamma_c^{sm})$, instead of the space $H^{\frac{1}{2},N}(\Gamma_c^s)$.

Let every polygonal domain Ω_h^ι, $\iota = 1,\ldots,r$, be covered by a triangulation $\mathcal{T}_{h,\Omega^\iota}$ in such a way that on the contact boundaries Γ_{ch}^{sm} the points of Γ_{ch}^s and Γ_{ch}^m are not identical, therefore, the mesh sizes $h_s \neq h_m$ and the global mesh size h is $h = \max\{h_s, h_m\}$.

Let us introduce the discrete approximation of the Lagrange multiplier space $M_{hH} = M_{hn} \times M_{Ht}$, where

Case I:

$$\begin{aligned}
W_{hH}(\cup_s\Gamma_{ch}^s) =& W_{hn}(\cup_s\Gamma_{ch}^s) \times W_{Ht}(\cup_s\Gamma_{ch}^s) = \\
&= \left\{\mathbf{v}_h^s \cdot \mathbf{n}^s|_{\cup_s\Gamma_{ch}^s}, \mathbf{v}_h \in V_h\right\} \times \left\{\mathbf{v}_h^s \cdot \mathbf{t}^s|_{\cup_s\Gamma_{ch}^s}, \mathbf{v}_h \in V_h\right\}, \\
M_{hn} =& \Big\{\mu_{hn} \in W_{hn}(\cup_s\Gamma_{ch}^s), \int_{\Gamma_c^s} \mu_{hn}\psi_h ds \geq 0, \\
&\forall\psi_h \in W_{hn},\ \psi_n \geq 0 \text{ a.e. on every } \Gamma_{ch}^s\Big\},
\end{aligned}$$

$$\begin{aligned}
M_{Ht} =& \Big\{\mu_{Ht} \in W_{Ht}(\cup_s\Gamma_{ch}^s), \int_{\Gamma_{ch}^s} \mu_{Ht}\psi_H ds - \int_{\Gamma_{ch}^s} g_{ch}^{sm}|\psi_H| ds \leq 0, \\
&\forall\psi_H \in W_{Ht}(\cup_s\Gamma_{ch}^s)\Big\},
\end{aligned}$$

Case II:

$$\begin{aligned}
W_{hH}(\cup_s\Gamma_{ch}^s) =& W_{hn}(\cup_s\Gamma_{ch}^s) \times W_{Ht}(\cup_s\Gamma_{ch}^s) = \\
&= \left\{\boldsymbol{\mu}_{hH}|_{\Delta_r} \in [P_0(\Delta_r)]^N,\ 0 \leq r \leq n(h^s)\right\},
\end{aligned}$$

$$\begin{aligned}
M_{hn} =& \{\mu_{hn} \in W_{hn}(\cup_s\Gamma_{ch}^s), \mu_{hn} \geq 0 \text{ a.e. on every } \Gamma_{ch}^s\}, \\
M_{Ht} =& \Big\{\boldsymbol{\mu}_{Ht} \in W_{Ht}(\cup_s\Gamma_{ch}^s), \int_{\Gamma_{ch}^s} \boldsymbol{\mu}_{Ht}\boldsymbol{\psi}_H ds - \int_{\Gamma_{ch}^s} g_{ch}^{sm}|\boldsymbol{\psi}_H| ds \leq 0, \\
&\forall\boldsymbol{\psi}_H \in W_{Ht}(\cup_s\Gamma_{ch}^s)\Big\},
\end{aligned}$$

where $\Gamma_{ch}^s = \cup_{p=1}^{\overline{p}} \Gamma_{ch}^{sp}$, $\Gamma_{ch}^{sp} = \cup_r \Delta_r$, $\Delta_r = (q_r, q_r + h_p)$, $r = 0,1,\ldots,m-1$, $m = m(h_p)$, where h_p is a step of mesh in the $p-th$ segment of Γ_{ch}^s for $N = 2$; or $\Gamma_{ch}^s = \cup_r \Delta_r$, where Δ_r are the faces of polyhedra, $r = 0,1,\ldots,m-1$, $m = m(h_p)$, for $N = 3$. Let

$$\begin{aligned}
b(\boldsymbol{\mu}_{hH}, \mathbf{v}_h) = \langle \mu_{hn}, [\mathbf{v}_h \cdot \mathbf{n}]^s - d^{sm} \rangle_{\cup\Gamma_{ch}^s} + \int_{\cup_s\Gamma_{ch}^s} g_{ch}^{sm} \boldsymbol{\mu}_{Ht} \cdot [\mathbf{v}_{ht}]^s ds, \\
\boldsymbol{\mu}_{hH} \in M_{hH},\ \mathbf{v}_h \in V_{0h},
\end{aligned}$$

where $[\mathbf{v}_h \cdot \mathbf{n}]^{sm} = v^s_{hn}(\mathbf{x},t) - v^m_{hn}(\mathcal{R}^{sm}(\mathbf{x},t))$, $[\mathbf{v}_{ht}]^{sm} = \mathbf{v}^s_{ht}(\mathbf{x},t) - \mathbf{v}^m_{ht}(\mathcal{R}^{sm}(\mathbf{x},t))$, where $\mathcal{R}^{sm}$: $\Gamma^s_{ch}(t) \mapsto \Gamma^m_{ch}(t)$, at $t \in I$, is a bijective map satisfying $\Gamma^m_{ch}(t) \subset \mathcal{R}^{sm}(\Gamma^s_{ch}(t)), t \in I$, and where $\langle \cdot, \cdot \rangle_{\Gamma^s_{ch}}$ denotes the duality pairing between W_{hH} and M_{hH}.

Remark 5. *A linear mapping $T : V \to W$ is called injective iff whenever $T(v_1) = T(v_2)$, we have $v_1 = v_2$. The mapping T is surjective iff $\forall w \in W, \exists v \in V : T(v) = w$. Moreover, T is bijective if it is injective and surjective.*

For the thermal part a mortar finite element approach can be used (McDevitt and Laursen (2000), Falletta and Lamichhane (2009)). Then we introduce the following spaces: $X = \prod_{\iota=1}^r H^1(\Omega^\iota)$, X_h is its finite element approximation, $X_h \subset X$, $M_1 := H^{\frac{1}{2}}(\cup_{sm}\Gamma^{sm}_c)$ the trace space on the common interfaces, $M_2 := H^{-\frac{1}{2}}(\cup_{sm}\Gamma^{sm}_c)$ its dual, and $M_{1h} \subset M_1$ and $M_{2h} \subset M_2$ are their discrete approximations. Then for every time level we will introduce the generalized saddle point problem. Then we introduce the bilinear forms $A^{(T)} : X \times X \to \mathbb{R}$, $b_1 : X \times M_1 \to \mathbb{R}$, $b_2 : X \times M_2 \to \mathbb{R}$, $c : M_1 \times M_2 \to \mathbb{R}$ and $g_f : M_2 \to \mathbb{R}$ and their discrete approximations. Furthermore, we introduce the heat flux between both neighbouring lithospheric bodies (i.e., from Ω^m to Ω^s and $\lambda_0 := \kappa^m \nabla \overline{T}^m.\mathbf{n}^m \in M_1$), and then a Lagrange multiplier follows. We recall that the inf-sup condition for the bilinear form $b_1(.,.)$ will be used. The corresponding algorithm we denote as Algorithm $(\mathcal{A}_T)$. The problem is modification of the method given in Falletta and Lamichhane (2009).

Then we have the corresponding problem:

Problem $(\mathcal{P})_h$**:** At every time level find $(\boldsymbol{\lambda}_{hH}, \mathbf{U}_h) \in M_{hH} \times V_h$ and $(\xi_h, \overline{T}_h) \in M_{2h} \times X_h$ satisfying

$$
\begin{aligned}
& A(\mathbf{U}_h, \mathbf{v}_h) + b(\boldsymbol{\lambda}_{hH}, \mathbf{v}_h) = (\mathbf{f}_h, \mathbf{v}_h) && \forall \mathbf{v}_h \in V_h = \prod_{\iota=1}^r V_h^\iota, \\
& b(\boldsymbol{\mu}_{hH} - \boldsymbol{\lambda}_{hH}, \mathbf{v}_h) \leq \langle d^{sm}, \mu_{hn} - \lambda_{hn} \rangle_{\cup \Gamma^s_{ch}} && \forall \boldsymbol{\mu}_{hH} \in M_{hH}, \qquad (60)
\end{aligned}
$$

$$
\begin{aligned}
& A^{(T)}(\overline{T}_h, z_h) + b_1(z_h, \lambda_{h0}) = (Q_h, z_h) && \forall z_h \in X_h = \prod_{\iota=1}^r X_h^\iota, \\
& b_2(\overline{T}_h, \xi_h) - c(\lambda_{h0}, \xi_h) = (g_f, \xi_h), && \forall \xi_h \in M_{2h}. \qquad (61)
\end{aligned}
$$

It can be shown that when $-\tau_n(\mathbf{U}_h) \in M_{hn}$ the above visco-elastic part of the problem has a unique solution $(\boldsymbol{\lambda}_{hH}, \mathbf{U}_h) \in M_{hH} \times V_h$, a.e. $t \in I$. Moreover, we have

$$\lambda^s_{hn} = -\tau^s_n(\mathbf{U}_h) \quad \text{and} \quad g^s_c \boldsymbol{\lambda}^s_{Ht} = -\boldsymbol{\tau}^s_t(\mathbf{U}_h),$$

where $\mathbf{U}_h$ is the solution of the discrete primal problem and $g^s_c \equiv g^{sm}_{ch}$.

Remark 6. *For the existence and uniqueness of the solution to (60) it is necessary to ensure that $\{\boldsymbol{\mu}_{hH} \in M_{hH}, b(\boldsymbol{\mu}_{hH}, \mathbf{v}_h) = 0, \forall \mathbf{v}_h \in V_h\} = \{\mathbf{0}\}$, which for the Case I is obvious and for the Case II the technique of Haslinger et al. (1996), with using the discrete Babuška-Brezzi "inf-sup" condition, is used.*

Further, in this section we limit ourselves onto the elastic part of the problem only.

Next, the primal-dual active set (PDAS) method for the elastic part of the problem will be introduced. The method will be parallel to that of Nedoma (2010), (2012a,b).

As usual in the mortar approach the interface Γ_{ch}^{sm} has two sides, the "slave" side from Ω_h^s and the "master" side from Ω_h^m. The interface Γ_{ch}^{sm} is assumed to be a union of faces in the 3D case. Similarly as Wohlmuth and Krause (2003) we introduce the modification of the space $V_h = \prod_{s,m}(V_h^s \times V_h^m) \subset V$ by such a way that the nodal basis function on the mortar side will be bi-orthogonal with respect to the piecewise linear basis on the slave side. Thus the non-penetration condition $[\mathbf{U}\cdot\mathbf{n}]^s \leq d^{sm}$ will be replaced by its weak discrete form

$$\int_{\cup\Gamma_{ch}^s} [\mathbf{U}_h\cdot\mathbf{n}]^s \varphi_p ds \leq \int_{\cup\Gamma_{ch}^s} d_p^s \varphi_p ds, \quad p \in S, \tag{62}$$

giving a coupling between the vertices on the slave side and the master side, where S is the set of all vertices in the potential contact part on the slave side. We introduce a basis transformation of the basis of V_h in such a way that the weak non-penetration condition (62) in the new basis (the so-called dual basis) only deals with the vertices on the slave side.

In the mortar approach, the Lagrange multiplier space is approximated by its $(N-1)$-dimensional mesh resulting from the N-dimensional triangulation on the slave side. Here we will use discontinuous piecewise linear nodal basis functions for the dual Lagrange multiplier. The discrete Lagrange multiplier space M_{hH} can be spanned as $M_{hH}^s = \text{span}\{\psi_i\mathbf{e}_k, i = 1,\ldots,n_c, k = 1,\ldots,N\}$, $s \in \{1,\ldots,r\}$, where ψ_i is the i-th scalar dual basis function, $\mathbf{e}_k$ is the k-th unit vector, n_c is the number of nodes on the slave side $\overline{\Gamma}_{ch}^s$, i.e., the number of freedom of the space M_{hH} in each component. Denoting $\{\varphi_j\}$, $j = 1,\ldots,m$, the standard piecewise linear basis on the slave side (i.e., the basis of a component $W_{hH}(\Gamma_c^{sm})$), we have

$$\int_{\Gamma_{ch}^s} \varphi_j \psi_i ds = \delta_{ij} \int_{\Gamma_{ch}^s} \varphi_j ds, \quad 1 \leq i,j \leq m, \tag{63}$$

where δ_{ij} is the Kronecker delta. Let W_{hH}^s be the vector valued trace space of V_{0h} restricted to $\cup_s\Gamma_{ch}^s$. Then for each $\mathbf{v}_h = \sum_i \gamma_i\varphi_i \in W_{hH}$ the discrete scalar product $\mathbf{v}_h\cdot\mathbf{n}_h = \sum_i(\gamma_i\cdot\mathbf{n}_i)\varphi_i$, where $\mathbf{n}_i$ denotes the outer normal of Ω^s at the node i. Similarly, for each $\boldsymbol{\mu}_h = \sum_i \boldsymbol{\alpha}_i\psi_i \in M_{hH}$ the discrete product $\boldsymbol{\mu}_h\cdot\mathbf{n}_h = \sum_i(\boldsymbol{\alpha}_i\cdot\mathbf{n}_i)\psi_i$. Let us define

$$\begin{aligned} W_h^{s+} &:= \{\mathbf{v}_h \in W_{hH}^s | \mathbf{v}_h\cdot\mathbf{n}_h \geq 0\}, \\ M_{hH}^{s+} &:= \left\{\boldsymbol{\mu}_{hH} \in M_{hH}^s | \langle\boldsymbol{\mu}_{hH},\mathbf{v}_h\rangle \geq 0, \mathbf{v}_h \in W_h^{s+}\right\}, \end{aligned}$$

and where

$$M_{hH}^s := \{\psi_i\mathbf{e}_k, i = 1,\cdots,m;\ k = 1,2,3\};\ s \in \{1,\cdots,r\},$$

where $\mathbf{e}_k$ are components of the unit Cartesian basis and

$$\begin{aligned} M_{hH}^+ &= \prod_s M_{hH}^{s+}, \\ b(\boldsymbol{\mu}_{hH},\mathbf{v}_h) &= \langle\boldsymbol{\mu}_{hH},[\mathbf{v}_h]^s\rangle_{\cup\Gamma_{ch}^s}. \end{aligned}$$

Moreover, let for completeness the discrete convex subset $K_h \subset V_h$ be as follows

$$K_h := \left\{\mathbf{v}_h \in V_h | b(\boldsymbol{\mu}_{hH},\mathbf{v}_h) \leq \int_{\Gamma_{ch}^s} d_h^{sm}(\boldsymbol{\mu}_{hH}\cdot\mathbf{n}_h)ds,\ \boldsymbol{\mu}_{hH} \in M_{hH}^+\right\},$$

where d_h^{sm} is a suitable approximation of d^{sm} on W_{hH}.

Then the discrete mortar formulation of the saddle point problem for every time level will be introduced as follows:

Problem $(\mathcal{P}_{sp})_{dm}$**:** At every time level find $\mathbf{U}_h \in V_h$, $\boldsymbol{\lambda}_{hH} \in M^+_{hH}$, a.e. $t \in I$, $\boldsymbol{\lambda}_{hH} = (\lambda_{hn}, \boldsymbol{\lambda}_{Ht})$, satisfying

$$\begin{aligned} A(\mathbf{U}_h, \mathbf{v}_h) + b(\boldsymbol{\lambda}_{hH}, \mathbf{v}_h) &= (\mathbf{f}_h, \mathbf{v}_h) \quad \forall \mathbf{v}_h \in V_h,\ t \in I, \\ b(\boldsymbol{\mu}_{hH} - \boldsymbol{\lambda}_{hH}, \mathbf{v}_h) &\leq \langle (\boldsymbol{\mu}_{hH} - \boldsymbol{\lambda}_{hH}) \cdot \mathbf{n}_h, d^{sm} \rangle_{\cup \Gamma^s_{ch}} \quad \forall \boldsymbol{\mu}_{hH} \in M^+_{hH},\ t \in I, \end{aligned} \tag{64}$$

where

$$\begin{aligned} A(\mathbf{U}_h, \mathbf{v}_h) =& (\mathbf{U}_h, \mathbf{v}_h) + \Delta t^2 a^{(e)}(\mathbf{U}_h; \mathbf{U}_h, \mathbf{v}_h) + \Delta t a^{(v)}(\mathbf{U}_h; \mathbf{U}_h, \mathbf{v}_h) \\ b(\boldsymbol{\mu}_{hH}, \mathbf{v}_h) =& \langle \boldsymbol{\mu}_{hH}, [\mathbf{v}_h]^s \rangle_{\cup \Gamma^s_{ch}} \quad \forall \mathbf{v}_h \in V_h,\ \boldsymbol{\mu}_{hH} \in M_{hH}\,. \end{aligned}$$

The Lagrange multipliers λ_{hn} are the approximations of the contact forces $-\boldsymbol{\tau}(\mathbf{U}_h) \cdot \mathbf{n} = -\left(n_j n_k \tau_{jk}(\mathbf{U}_h)\right)$ which are necessary to adjust the contact displacements on the contact boundary $\cup \Gamma^{sm}_{ch}$.

Remark 7. *For the frictionless case the tangential stress component on the contact boundary* $\cup \Gamma^{sm}_c$ *will be zero, then* $\boldsymbol{\lambda}_{Ht} = \mathbb{0}$, *and* $M_{hH} = M_{hn}$.

Let us denote the nodal parameters $\mathbf{U}_h$ by $\hat{\mathbf{U}}_h$ and of λ_{hH} by $\boldsymbol{\Lambda}_{hH}$.

The Eq. (64a) for every $t \in I$ in the matrix form is of the form

$$\mathbb{A}_h \hat{\mathbf{U}}_h + \mathbb{B}_h \boldsymbol{\Lambda}_{hH} = \mathbb{F}_h\,. \tag{65}$$

Now, when we can to determine the structure of $\mathbb{B}_h$, we decompose the set of all vertices in every time level into three disjoint parts $\mathcal{N}$, $\mathcal{M}$ and $\mathcal{S}$, where $\mathcal{S}(\dim n_v)$ denotes all vertices on the possible contact part on the slave side, $\mathcal{M}(\dim n_{cn})$ all vertices of the possible contact part of the master side, and $\mathcal{N}$ all the other one. Then the strong formulation of the non-penetration condition will be replaced by its weak discrete form

$$\int_{\cup \Gamma^{sm}_{ch}} [\mathbf{U}_h \cdot \mathbf{n}]^s \psi_p ds \leq \int_{\cup \Gamma^{sm}_{ch}} d^s_h \psi_p ds, \quad p \in \mathcal{S}, \tag{66}$$

that connects points of sets $\mathcal{S}$ and $\mathcal{M}$.

As usual in the discrete mortar techniques we introduce the useful transformations, then after some modifications (see e.g., Hintermüller et al. (2003), (2004), Brunsson et al. (2007), Nedoma (2010), (2012a,b)), we have to solve in every time level the following problem

$$\begin{aligned} &\hat{\mathbb{A}}_h \hat{\mathbb{U}} + \hat{\mathbb{B}}_h \boldsymbol{\Lambda}_{hH} = \hat{\mathbb{F}}_h, \\ &\hat{\mathbb{U}}_{n,p} \leq d^{sm}_p, \boldsymbol{\Lambda}_{hn,p} \geq 0, (\hat{\mathbb{U}}_{n,p} - d^{sm}_p) \boldsymbol{\Lambda}_{hn,p} = 0, \quad \forall p \in \mathcal{S},\ t \in I, \\ &\boldsymbol{\Lambda}_{Ht,p} = (\boldsymbol{\Lambda}_{hn,p}, \boldsymbol{\Lambda}_{Ht,p}), \quad \forall p \in \mathcal{S},\ t \in I, \end{aligned} \tag{67}$$

where the condition (b),(c) and (d) represent the Karush-Kuhn-Tucker conditions of a constrained optimization problem for inequality constraints, and where

$$\begin{aligned} \boldsymbol{\Lambda}_{hn,p} &= \mathbf{n}^{sT}_p \hat{\mathbb{D}}[p,p] \boldsymbol{\Lambda}_{hH}(p), \quad \boldsymbol{\Lambda}_{hH}(p) \in \mathbb{R}^N, \\ \boldsymbol{\Lambda}_{Ht,p} &= \boldsymbol{\Lambda}_{hH}(p) - (\boldsymbol{\Lambda}_{hH}(p) \cdot \mathbf{n}^s_p) \mathbf{n}^s_p = (\boldsymbol{\Lambda}_{hH}(p) \cdot \mathbf{t}^s_p) \mathbf{t}^s_p, \end{aligned} \tag{68}$$

and where the matrix $\hat{\mathbb{B}}_h = Q\mathbb{B}_h = (\mathbb{O},\mathbb{O},\mathbb{D})^T$, $Q = \begin{bmatrix} \mathbb{I}_N & \mathbb{O} & \mathbb{O} \\ \mathbb{O} & \mathbb{I}_N & \hat{\mathbb{M}}^T \\ \mathbb{O} & \mathbb{O} & \mathbb{I}_N \end{bmatrix}$, the matrix $\mathbb{D} = (\mathbb{D}[p,q])$, where $\mathbb{D}[p,q]) = \delta_{pq}\mathbb{I}_N \cdot \int_{\cup\Gamma_c^{sm}} \varphi_p \psi_p ds, p \in \mathcal{S}$, is the diagonal matrix after the used transformations and where $\hat{\mathbb{M}} = \mathbb{D}^{-1}\mathbb{M}$, $\mathbb{M} = (\mathbb{M}[p,q])$, $\mathbb{M}[p,q] = \mathbb{I}_N \int_{\cup\Gamma_c^{sm}} \varphi_p \psi_p ds, p \in \mathcal{S}, q \in \mathcal{M}$, is the block matrix representing coupling between the traces of the finite element shape functions on the master side "m" and the shape functions for the Lagrange multiplier on the slave side "s" (for details see Nedoma (2010), (2012a,b)). Moreover, the matrix $\hat{\mathbb{A}}$ is of the form

$$\hat{\mathbb{A}}_h = Q\mathbb{A}_h Q^T =$$
$$= \begin{bmatrix} \mathbb{A}_{\mathcal{N}\mathcal{N}} & \mathbb{A}_{\mathcal{N}\mathcal{M}} + \mathbb{A}_{\mathcal{N}\mathcal{S}}\hat{\mathbb{M}} & \mathbb{A}_{\mathcal{N}\mathcal{S}} \\ \mathbb{A}_{\mathcal{M}\mathcal{N}} + \hat{\mathbb{M}}^T\mathbb{A}_{\mathcal{S}\mathcal{N}} & \begin{matrix} \mathbb{A}_{\mathcal{M}\mathcal{M}} + \mathbb{A}_{\mathcal{M}\mathcal{S}}\hat{\mathbb{M}} + \\ +\hat{\mathbb{M}}^T\mathbb{A}_{\mathcal{S}\mathcal{M}} + \hat{\mathbb{M}}^T\mathbb{A}_{\mathcal{S}\mathcal{S}}\hat{\mathbb{M}} \end{matrix} & \mathbb{A}_{\mathcal{M}\mathcal{S}} + \hat{\mathbb{M}}^T\mathbb{A}_{\mathcal{S}\mathcal{S}} \\ \mathbb{A}_{\mathcal{S}\mathcal{N}} & \mathbb{A}_{\mathcal{S}\mathcal{M}} + \mathbb{A}_{\mathcal{S}\mathcal{S}}\hat{\mathbb{M}} & \mathbb{A}_{\mathcal{S}\mathcal{S}} \end{bmatrix},$$

and the vector $\hat{\mathbb{F}}_h$ is of the form

$$\hat{\mathbb{F}}_h = Q\mathbb{F}_h = (\mathbb{F}_{\mathcal{N}}, \mathbb{F}_{\mathcal{M}} + \hat{\mathbb{M}}^T\mathbb{F}_{\mathcal{S}}, \mathbb{F}_{\mathcal{S}})^T .$$

The weak non-penetration condition after the used transformation is of the form (see Nedoma (2010), (2012a,b))

$$\hat{\mathbb{U}}_{n,p} \equiv (\mathbf{n}_p^s)^T\mathbb{D}[p,p]\hat{\mathbb{U}}_p \leq d_p^{sm} \quad \forall p \in \mathcal{S}, \tag{69}$$

where $d_p^{sm} = \int_{\cup_s\Gamma_c^s} d_h^{sm}\psi_p ds$, $p \in \mathcal{S}$, as the coefficients at $\hat{\mathbb{U}}_q$, $q \in \mathcal{M}$, are nullified.

As usual in the primal-dual active set (PDAS) algorithm the decomposition of the set of vertices $\mathcal{S}$ on the slave side in each step of the PDAS algorithm into the disjoint active and inactive sets as $\mathcal{S} = \mathcal{A}_k \cup I_k$ is used.

Let in the 3D model with the Tresca friction the displacements $\hat{\mathbf{U}}_h$ be known, the Lagrange multiplier $\mathbf{\Lambda}_{hH} = (\Lambda_{hn}, \mathbf{\Lambda}_{Ht})$ can be computed directly from (65), that is,

$$\mathbf{\Lambda}_{hH} = \mathbb{D}^{-1}(\mathbb{F}_h - \mathbb{A}_h\hat{\mathbf{U}}_h)_S , \tag{70}$$

where the subscript S denotes that we use only the entries of the vector corresponding to the nodes $p \in S$. For the normal and tangential components of the multiplier $\mathbf{\Lambda}_{hH}$ and of the relative decomposition $\hat{\mathbf{U}}_h$ for a node point $p \in S$, we have

$$\begin{aligned} \hat{U}_{hn,p} &= \hat{\mathbf{U}}_{hp}^T\mathbf{n}_p \in \mathbb{R}, \hat{\mathbf{U}}_{Ht,p} = (\hat{\mathbf{U}}_{hp}^T\mathbf{t}_{1p}, \hat{\mathbf{U}}_{hp}^T\mathbf{t}_{2p})^T \in \mathbb{R}^2, \\ \mathbf{\Lambda}_{hn,p}^s &= (\mathbf{n}_p^s)^T\mathbb{D}[p,p]\mathbf{\Lambda}_{hH}(p) \in \mathbb{R}, \quad \mathbf{\Lambda}_{hH}(p) \in \mathbb{R}^3, \\ \mathbf{\Lambda}_{Ht,p}^s &= \mathbf{\Lambda}_{hH}(p) - (\mathbf{\Lambda}_{hH}(p)\cdot\mathbf{n}_p^s)\mathbf{n}_p^s = \left(\mathbf{\Lambda}_{hH}(p)\cdot\mathbf{t}_p^s\right)\mathbf{t}_p^s = \\ &= \left(\mathbf{\Lambda}_{hH}^T(p)\mathbb{D}[p,p]\mathbf{t}_{1,p}^s, \quad \mathbf{\Lambda}_{hH}^T(p)\mathbb{D}[p,p]\mathbf{t}_{2,p}^s\right)^T \in \mathbb{R}^2 . \end{aligned}$$

Let us put

$$C_n(\hat{\mathbb{U}}_{n,p}, \mathbf{\Lambda}_{hn,p}) = \mathbf{\Lambda}_{hn,p} - \max\left\{0, \mathbf{\Lambda}_{hn,p} + c_1(\hat{\mathbb{U}}_{n,p} - d_p^{sm})\right\}, \quad c_1 = \text{const.} > 0,$$

then (67b–d) can be expressed as

$$C_n(\hat{\mathbb{U}}_{n,p}, \mathbf{\Lambda}_{hn,p}) = 0, \quad p \in \mathcal{S}.$$

Then the discrete conditions for normal contact are given by (67b,c,d) and the discrete Coulomb friction conditions are given by

$$\begin{aligned} &\left|\mathbf{\Lambda}_{Ht,p}^s(p)\right| \leq \mathcal{F}_c^{sm}\left|\mathbf{\Lambda}_{hn,p}^s\right| \\ &\left|\mathbf{\Lambda}_{Ht,p}^s(p)\right| < \mathcal{F}_c^{sm}\left|\mathbf{\Lambda}_{hn,p}^s\right| \Rightarrow \hat{\mathbf{U}}_{Ht,p} = \mathbf{0} \\ &\left|\mathbf{\Lambda}_{Ht,p}^s(p)\right| = \mathcal{F}_c^{sm}\left|\mathbf{\Lambda}_{hn,p}^s\right| \Rightarrow \exists\vartheta \geq 0 \text{ such} \\ &\qquad\qquad\qquad\qquad\qquad \text{that } \mathbf{\Lambda}_{Ht,p}^s = -\vartheta\hat{\mathbf{U}}_{Ht,p} \end{aligned} \qquad \text{for all } p \in S, \tag{71}$$

where for the Tresca friction model $\mathcal{F}_c^{sm}|\mathbf{\Lambda}_{hn,p}^s| \equiv g_{ch,p}^s \in H^{-\frac{1}{2}}(\Gamma_c^s)$, $g_{ch,p}^s \geq 0$, $g_{ch,p}^s = \int_{\Gamma_{ch}^s} g_{ch}^s \varphi_p ds$. For $g_{ch,p}^s = 0$ the condition (71) leads to homogeneous Neumann boundary conditions in tangential direction. Let $g_{ch,p}^s > 0$, then the condition (71) is equivalent to $C_t\left(\hat{\mathbb{U}}_{t,p}, \mathbf{\Lambda}_{Ht,p}^s\right) = 0$ for all $p \in S$, where

$$\begin{aligned} &C_t\left(\hat{\mathbb{U}}_{t,p}, \mathbf{\Lambda}_{Ht,p}^s\right) = \\ &= \max\left(g_{ch,p}^s, |\mathbf{\Lambda}_{Ht,p}^s + c_2\hat{\mathbb{U}}_{t,p}|\right)\mathbf{\Lambda}_{Ht,p}^s - g_{ch,p}^s\left(\mathbf{\Lambda}_{Ht,p}^s + c_2\hat{\mathbb{U}}_{t,p}\right), \quad c_2 > 0, \end{aligned} \tag{72}$$

which will be the starting point of the algorithm, that will be based on a Newton-type algorithm for the solution $C_t(\hat{\mathbb{U}}_{t,p}, \mathbf{\Lambda}_{Ht,p}^s) = 0$. As it was shown in Hintermüller et al. (2003) the max-function and the Euclidean norm are semi-smooth, and therefore, a semi-smooth Newton method can be used. If the Euclidean norm $\left|\mathbf{\Lambda}_{Ht,p}^s + c_2\hat{\mathbb{U}}_{t,p}\right| = 0$, then $\max\left(g_{ch,p}^s, |\mathbf{\Lambda}_{Ht,p}^s + c_2\hat{\mathbb{U}}_{t,p}|\right) = g_{ch,p}^s$ and the Euclidean norm vanishes. Hence, the derivative of the Euclidean norm only occurs for points that are differentiable in the classical sense. The analysis of the generalized derivative of $C_t(\hat{\mathbb{U}}_{t,p}, \mathbf{\Lambda}_{Ht,p}^s)$ (see Hüeber et al. (2008)) shows that the nodes of $\mathcal{S}$ are separated into the active set $\mathcal{A}_{Ht,k}$ and the inactive set $I_{Ht,k}$ defined by

$$\mathcal{A}_{Ht,k} := \left\{p \in S; |\mathbf{\Lambda}_{Ht,p}^{s,k-1} + c_2\hat{\mathbb{U}}_{t,p}^{k-1}| - g_{ch,p}^s > 0\right\}, \tag{73}$$

$$I_{Ht,k} := \left\{p \in S; |\mathbf{\Lambda}_{Ht,p}^{s,k-1} + c_2\hat{\mathbb{U}}_{t,p}^{k-1}| - g_{ch,p}^s \leq 0\right\}. \tag{74}$$

Separation of nodes of S into the active set $\mathcal{A}_{Ht,k}$ and the inactive set $I_{Ht,k}$ is provided by the characteristic function in the generalized derivative of $C_t(\cdot,\cdot)$. The cases if $g_{ch,p} = 0$ and $g_{ch,p} > 0$ are studied in Hüeber et al. (2008).

Summing all results, then Problem $(\mathcal{P}_h)$ can be rewritten as

$$\begin{aligned} \hat{\mathbb{A}}_h\hat{\mathbb{U}}+\hat{\mathbb{B}}_h\mathbf{\Lambda}_{hH} &= \hat{\mathbb{F}}_h, \\ C_n\left(\hat{\mathbb{U}}_{n,p},\mathbf{\Lambda}_{hn,p}\right) &= 0, \\ C_t(\hat{\mathbb{U}}_{t,p}\mathbf{\Lambda}_{Ht,p}) &= 0 \end{aligned} \tag{75}$$

for all vertices $p \in \mathcal{S}$ and $t \in I$.

The algorithm of the Tresca contact model in the thermo-visco-elastic rheology has two main steps $M1$ and $M2$. The Step $M1$ is represented by the Algorithm$(\mathcal{T})$ and the Step $M2$ is represented by the algorithm solving the linear system (61) at every time level (Falleta and Lamichhame (2009)).

The PDAS algorithm for the contact problem with friction in the Tresca sense is the following:

The Main Step M1:

Algorithm ($\mathcal{T}$):

STEP 1: Initiate the active sets $\mathcal{A}_{hn,1}$, $\mathcal{A}_{Ht,1}$ and the inactive sets $I_{hn,1}$, $I_{Ht,1}$ such that $\mathcal{S}_n = \mathcal{A}_{hn1} \cup I_{hn1}$, $\mathcal{S}_t = \mathcal{A}_{Ht1} \cup I_{Ht1}$, $\mathcal{A}_{hn1} \cap I_{hn1} = \emptyset$, $\mathcal{A}_{Ht1} \cap I_{Ht1} = \emptyset$ and put the initial value $(\hat{\mathbb{U}}^0, \mathbf{\Lambda}_{hH}^0)$, $c_1, c_2 \in (10^3, 10^4)$ and set $k = 1$, $c_1 > 0$, $c_2 > 0$, $m \in \mathbb{N}$ and initiate the coupled terms.

STEP 2: Define the active and inactive sets

$$\begin{aligned} \mathcal{A}_{hn,k} &:= \left\{p \in S; \mathbf{\Lambda}_{hn,p}^{s,k-1} + c_1\left(\hat{\mathbb{U}}_{n,p}^{k-1,m} - d_p^{sm}\right) > 0\right\}, \\ I_{hn,k} &:= \left\{p \in S; \mathbf{\Lambda}_{hn,p}^{s,k-1} + c_1\left(\hat{\mathbb{U}}_{n,p}^{k-1,m} - d_p^{sm}\right) \leq 0\right\}, \\ \mathcal{A}_{Ht,k} &:= \left\{p \in S; \left|\mathbf{\Lambda}_{Ht,p}^{s,k-1} + c_2\hat{\mathbb{U}}_{t,p}^{k-1,m}\right| - g_{ch,p}^s > 0\right\}, \\ I_{Ht,k} &:= \left\{p \in S; \left|\mathbf{\Lambda}_{Ht,p}^{s,k-1} + c_2\hat{\mathbb{U}}_{t,p}^{k-1,m}\right| - g_{ch,p}^s \geq 0\right\}, \end{aligned}$$

and define the coupled terms with the values obtained by the Algorithm $\mathcal{A}_T$, and define the vector $\hat{\mathbb{F}}_{hS}$.

STEP 3: For $i = 1, \ldots, m$, compute the generalized derivative in the sense of a semi-smooth Newton method, i.e.,

$$\hat{\mathbb{U}}_{hH}^{k,i} = G\left(\hat{\mathbb{U}}_{hH}^{k,i-1}, \mathcal{A}_{hn,k}, I_{hn,k}, \mathcal{A}_{Ht,k}, I_{Ht,k}, \hat{\mathbb{U}}_{hH}^{k-1,m}, \mathbf{\Lambda}_{hH}^{k-1}\right),$$

where by the symbol G we denote the generalized derivative in the sense of a semi-smooth Newton method.

STEP 4: **If** $\left|\hat{\mathbb{U}}_{hH}^{k,m} - \hat{\mathbb{U}}_{hH}^{k,0}\right| / \left|\hat{\mathbb{U}}_{hH}^{k,m}\right| < \varepsilon$ **then** STOP.

STEP 5: Compute the Lagrange multiplier due to (70), that is,

$$\mathbf{\Lambda}_{hH}^k = \hat{\mathbb{D}}^{-1}\left(\hat{\mathbb{F}}_{hS} - \hat{\mathbb{A}}_{hS}\hat{\mathbb{U}}_{hH}^{k,m}\right).$$

STEP 6: Set $\hat{\mathbb{U}}_{hH}^{k+1,0} = \hat{\mathbb{U}}_{hH}^{k,m}$, $k = k+1$ and **goto STEP 2.**

The system (75) can be rewritten, if we decompose the set of vertices S on the slave side in each step k of the PDAS algorithm into the disjoint active and inactive sets (for more details see Hüeber et al. (2008), Nedoma (2010), (2012a,b)). Since due to the dual Lagrange multiplier space, $\mathbf{\Lambda}_{hH}^{k}$ can be eliminated locally.

The Step M2:

The Algorithm $(\mathcal{A}_T)$ solved in every time level $t = t_{i+1}$ the Eq. (61) will be based on the mortar finite element approach for a heat transfer problem (see Falletta and Lamichhane (2009); the algorithm of McDevitt and Laursen (2000) can be also used).

A primal-dual active set (PDAS) algorithm with the Coulomb friction

The algorithms can be based on the fixpoint algorithm or on the full Newton method (Hüeber et al. (2008), Nedoma (2012b)).

The Fixpoint Algorithm (FP) is the extension of the above PDAS algorithm for the Tresca friction, in which the friction bound $g_{ch,p}^{s} = \mathcal{F}_c^{sm}|\mathbf{\Lambda}_{n,p}^{s}|$ is iteratively modified using the normal component of the Lagrange multiplier. Thus, we have the following algorithm, that the friction bound and the active and inactive sets are updated in every step.

Algorithm (FP):

STEP 1: Initiate the initial value $(\hat{\mathbb{U}}^{0,0}, \mathbf{\Lambda}_{hH}^{0})$, $c_1, c_2 \in (10^3, 10^4)$ and set $k = 1, m \in \mathbb{N}$.

STEP 2: Set $k_c = k-1$ and update the friction bound by $g_{ch,p}^{s,k_c} = \mathcal{F}_c^{sm}\max\{0, \mathbf{\Lambda}_{n,p}^{s,k_c}\}$, $p \in S$.

STEP 3: Define the active sets $\mathcal{A}_{hn,k}$, $\mathcal{A}_{Ht,k}$ and the inactive sets $I_{hn,k}$, $I_{Ht,k}$ by

$$\mathcal{A}_{hn,k} := \left\{ p \in S; \mathbf{\Lambda}_{hn,p}^{s,k-1} + c_1\left(\hat{\mathbb{U}}_{n,p}^{k-1,m} - d_p^{sm}\right) > 0 \right\},$$
$$I_{hn,k} := \left\{ p \in S; \mathbf{\Lambda}_{hn,p}^{s,k-1} + c_1\left(\hat{\mathbb{U}}_{n,p}^{k-1,m} - d_p^{sm}\right) \leq 0 \right\},$$
$$\mathcal{A}_{Ht,k} := \left\{ p \in S; \left|\mathbf{\Lambda}_{Ht,p}^{s,k-1} + c_2\hat{\mathbb{U}}_{t,p}^{k-1,m}\right| - g_{ch,p}^{s,k_c} > 0 \right\},$$
$$I_{Ht,k} := \left\{ p \in S; \left|\mathbf{\Lambda}_{Ht,p}^{s,k-1} + c_2\hat{\mathbb{U}}_{t,p}^{k-1,m}\right| - g_{ch,p}^{s,k_c} \leq 0 \right\}.$$

STEP 4: For $i = 1, \ldots, m$, compute the generalized derivative in the sense of a semi-smooth Newton method

$$\hat{\mathbb{U}}_{hH}^{k,i} = G\left(\hat{\mathbb{U}}_{hH}^{k,i-1}, \mathcal{A}_{hn,k}, I_{hn,k}, \mathcal{A}_{Ht,k}, I_{Ht,k}, \hat{\mathbb{U}}_{hH}^{k-1,m}, \mathbf{\Lambda}_{hH}^{k-1}\right),$$

where the symbol G has the same meaning as above.

STEP 5: Compute the Lagrange multiplier due to (70) as

$$\mathbf{\Lambda}^k_{hH} = \hat{\mathbb{D}}^{-1}\left(\hat{\mathbf{F}}_{hS} - \hat{\mathbb{A}}_{hS}\hat{\mathbb{U}}^{k,m}_{hH}\right).$$

STEP 6: If $\left\|\hat{\mathbb{U}}^{k,m}_{hH} - \hat{\mathbb{U}}^{k_c,m}_{hH}\right\| / \left\|\hat{\mathbb{U}}^{k,m}_{hH}\right\| < \varepsilon$ **then** STOP.

STEP 7: Set $\hat{\mathbb{U}}^{k+1,0}_{hH} = \hat{\mathbb{U}}^{k,m}_{hH}$ and $k = k+1$ and **goto** STEP 2.

If $m = \infty$, we obtain the exact version of the algorithm, in the previous case we speak about inexact algorithm. The algorithm is convergent for the small coefficients of friction (see the estimate (32) and the estimates in Eck et al. (2005) for the continuous cases).

Remark 8. *The Full Newton method can be also used, because it is very advantageous method due to its fast convergence (for the derivation see e.g., in Hüeber et al. (2008)). This fast convergence is due to the fact that the friction bound is updated in the Newton iteration.*

The Step M2:

The algorithm is parallel to that introduced by Falleta and Lamichhane (2009) that we denote above as the Algorithm $(\mathcal{A}_T)$.

3.5. Opening of Deep Fractures and Faults. Numerical Algorithms

In the real situations it is very difficult to determine the location of a fracture (fault), its initiation, its further opening and propagation, and in which direction it propagates. In the literature many fracture initiation, fracture propagation and kinking criteria were introduced. These criteria depend on the material properties and also depend on the length of possible fractures. Some of them are based on critical values of stress and strain or on the use of a damage law and a critical damage parameter to determine the location of the fracture initiation.

To determine the areas of possible fracture zones, we firstly determine the areas with maximal principle stresses, and therefore, the places where fractures or faults are initiated (Figs 2a,b). Thus we need to check, at each time step, when the fault starts to propagate and in which direction. The algorithms will be based on the methods of the fracture mechanics. In the first case the fracture propagation criteria will be used, while in the second one the crack kinking criteria will be used. When a fault start to propagate through a mesh, the accuracy at the fracture tip will be of a great importance for determination of a possible fracture and a fault. Many numerical tools were developed to improve the accuracy at the fracture tip. Since the stress field is singular in the vicinity of the fracture tip, a concentric mesh around the fracture (fault) tip can be coupled with singular elements, that can be used to model the stress field singularity (Whiteman and Akin (1979), Nedoma (1998)). An other approach is based on the strain energy release rate, where a construction of ring elements in the neighborhood of the fracture and the fault tips can be also used. Finally mesh refinement around the fracture and the fault tips are necessary to keep a better precision in the vicinity of the fracture and the fault. Since the fracture and the fault propagate,

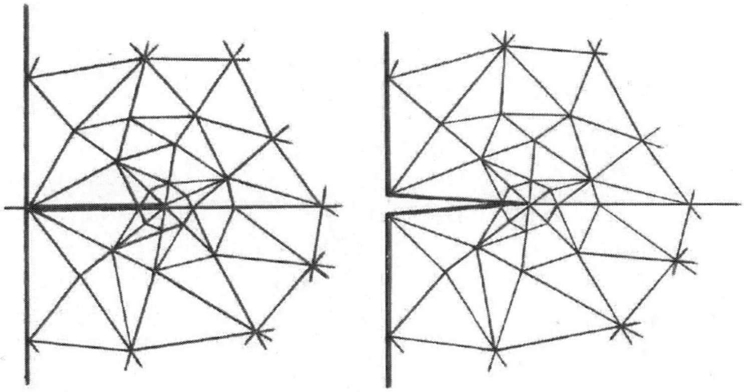

(a) (b)

Figure 2. Location of the deep fracture (or the deep fault) and the mesh around the fracture (or the fault) tip: (a) initiation state; (b) opening evolution state.

the fracture and the fault tips move along and the areas in the vicinity of fracture and the vicinity of fault are changed; thus, a new mesh is created and refined only in areas at the front of the propagated fracture and fault. For more details see Whiteman and Akin (1984), Nedoma (1998), Gdoutos (2005).

Many numerical algorithms have been applied to improve the accuracy at the fracture tip and to determine a fracture propagation direction. With a great advantage the automatic re-meshing procedure at the fracture tip, with a thickening of the mesh at the fracture tip and using singular elements to model the singular stress-strain fields, can be used. To determine a fracture propagation direction we compute the eigenvalues and the eigenvectors of the stress tensor in all determined mesh points nearest to the fracture tip, i.e., we determine the principal stresses and their directions. The final direction of the fracture propagation will be obtained as a weighted average of each direction with respect to the distance between the mesh point and the fracture tip. Moreover, stress intensity factors, that is, strength singularity at the fracture tip, can be used for determination of a fracture propagation.

With opening the fracture (fault) (Fig.2b), at the fracture (fault) faces, we denote them by ${}^{i}\Gamma_{c}^{s}$, $i = 1,2$, the contact conditions of Signorini type with or without Coulomb friction will be introduced. The numerical realization of such problem corresponds to formulation of a generalized dynamic fracture problem, formulated by using the non-penetration (Signorini type) conditions. Such problem leads to solve the dynamic contact problems and the PDAS algorithm that will be similar of that as discussed in the previous part of this section. Moreover, such problem leads to solve the dynamic contact problems by the Newmark method and the PDAS algorithms. It is evident that the evolutionary mesh will be created and refined only at the tip of the fractures and in the areas where it is needed in order to optimise calculations of the mesh and it is unrefined in the rest of the initial mesh.

4. Mathematical Model of Earthquakes in Regions with Possible Typhoon based on the Thermo-visco-plastic Rheology of Bingham Type

4.1. The Model

In this section we will discuss the other model that will facilitate to simulate the earthquake origin and ensuing tsunami, the effect of typhoon as well as the consequences of volcanic activity, and moreover, the submarine landslides and mudslides that may be evoked by the tsunamiearthquake and by its aftershocks, and that can evoke the ensuing further tsunamis. Similarly as in the previous model all these effects need to satisfy the main rheological properties in the investigated region. As we mentioned in the previous section the suitable choice of the rheology seems to be fundamental for the description of the earthquake mechanism, for all the effects concerning with the evolution of the tsunamis that can be strengthened by the effects of typhoon and the ensuing demanded security of the distressed continental areas. Then the thermo-visco-plastic rheology of Bingham type seems to be an useful rheology (Nedoma (2006a,b)).

Let the subduction zone (Fig.1) denoted by $\Omega \subset \mathbb{R}^N$, $N = 2,3$, with the boundary $\partial\Omega$, be a region occupied by a system of r geological bodies, that is, by a system of lithospheric plates and lithospheric blocks of arbitrary shapes Ω^ι, such that $\Omega = \cup_{\iota=1}^r \Omega^\iota$. Let all Ω^ι have Lipschitz boundaries $\partial\Omega^\iota$, let $\partial\Omega = \Gamma_\tau \cup \Gamma_u \cup \Gamma_c$, where the disjoint parts Γ_τ, Γ_u, Γ_c are open subsets. Let neighboring geological bodies, say Ω^s and Ω^m, are in a mutual contact. Denote the common contact boundary between both lithospheric plates or lithospheric blocks Ω^s and Ω^m before deformation by Γ_c^{sm}. Let $\mathbf{n} = (n_i)$ denote the unit outer normal vector and $\mathbf{t} = (t_i)$ denote the unit tangential vector to the boundary $\partial\Omega$ including the contact boundaries Γ_c^{sm}. Denote by $\mathbf{u} = (u_i)$ the velocity vector and by $\boldsymbol{\tau} = (\tau_i)$, $\tau_i = \tau_{ij}n_j$, $i, j = 1, \ldots, N$, the stress vector, its normal and tangential components by $\tau_n = \tau_i n_i = \tau_{ij}n_i n_j$, $\boldsymbol{\tau}_t = \boldsymbol{\tau} - \tau_n\mathbf{n}$, and the normal and tangential components of velocity vector by $u_n = u_i n_i$ and $\mathbf{u}_t = \mathbf{u} - u_n\mathbf{n}$. Denote by $\mathbf{u}^s, \mathbf{u}^m, T^s, T^m$ (indices s, m correspond with the neighboring geological bodies being in contact) the velocities and the temperatures in the neighboring geological bodies. All these quantities are functions of spatial coordinates and time. Let us assume that $\Gamma_u = {}^1\Gamma_u \cup {}^2\Gamma_u$, where similarly as in the previous section we set ${}^2\Gamma_u \equiv {}^2\Gamma_u \cup {}^3\Gamma_u \cup {}^4\Gamma_u$, $\Gamma_\tau = {}^1\Gamma_\tau \cup {}^2\Gamma_\tau$ and $\Gamma_c = \cup_{s,m}\Gamma_c^{sm}$, $\Gamma_c^{sm} = \partial\Omega^s \cap \partial\Omega^m$, $s \neq m$, $s, m \in \{1, \ldots, r\}$. Let $\Omega(t) = I \times \Omega$ be the time-space domain and by $\Gamma_\tau(t) = \Gamma_\tau \times I$, $\Gamma_u(t) = \Gamma_u \times I$, $\Gamma_c(t) = \Gamma_c \times I$ the parts of its boundary $\partial\Omega(t) = \partial\Omega \times I$. Similarly as in the previous section the influence of rotation of the Earth will be neglected.

Then we have to solve the following model problem:

Problem $(\mathcal{P}_{sm}^B)$: Let $N = 2,3$, $r \geq 2$. Find a pair of functions $(\mathbf{u}, T) : \Omega \times I \to (\mathbb{R}^N \times \mathbb{R}) \times I$,

$N=2,3,,$ satisfying

$$\rho^\iota\left(\frac{\partial u_i^\iota}{\partial t}+u_k\frac{\partial u_i^\iota}{\partial x_k}\right)+\alpha^\iota\frac{\partial u_i^\iota}{\partial t}=\frac{\partial \tau_{ij}^\iota(\mathbf{u}^\iota,T^\iota)}{\partial x_j}+F_i^\iota, \tag{76}$$

$$div\,\mathbf{u}=0,$$

$$\rho^\iota c_e^\iota\frac{\partial T^\iota}{\partial t}-\frac{\partial}{\partial x_j}\left(\kappa_{ij}^\iota\frac{\partial T^\iota}{\partial x_i}\right)=W^\iota-\rho^\iota\beta_{ij}^\iota T_0^\iota e_{ij}(\mathbf{u}^\iota)+$$
$$+c_{ijkl}^{(v)\iota}(\mathbf{u}^\iota)e_{kl}(\mathbf{u}^\iota)e_{ij}(\mathbf{u}^\iota), \tag{77}$$

where $i,j=1,\dots,N,\ \iota=1,\dots,r,\ (t,\mathbf{x})\in\Omega^\iota(t)=I\times\Omega^\iota,\ \iota=1,\dots,r,$

with the constitutive stress-strain relation the in thermo-Bingham rheology defined by

$$\tau_{ij}=\tau_{ij}^{(B)}+\tau_{ij}^{(T)}=-p\delta_{ij}+\hat{g}D_{ij}D_{II}^{-1/2}+2\hat{\mu}D_{ij}-\beta_{ij}(T-T_0)\,. \tag{78}$$

$$\text{provided } D_{II}\neq 0,\ D_{ij}=\frac{1}{2}\left(\frac{\partial u_i}{\partial x_j}+\frac{\partial u_j}{\partial x_i}\right),\ D_{II}=\frac{1}{2}D_{ij}D_{ij}, \tag{79}$$

the boundary value conditions

$$T(\mathbf{x},t)=T_1(\mathbf{x},t),\quad \tau_{ij}n_j=P_i,\quad i,j=1,\dots,N,,$$
$$\text{for } (t,\mathbf{x})\in\Gamma_\tau(t)=I\times\cup_{\iota=1}^r(\Gamma_\tau\cap\partial\Omega^\iota), \tag{80}$$
$$\kappa_{ij}\frac{\partial T(\mathbf{x},t)}{\partial x_i}n_j=K(Y(\mathbf{x})-T(\mathbf{x},t))\quad\left(\text{or } \kappa_{ij}\frac{\partial T(\mathbf{x},t)}{\partial x_i}n_j=q\right),$$
$$u_i=\ {}^1u_{2i}\equiv 0,\ i=1,\dots,N,\ \text{for } (t,\mathbf{x})\in\ {}^1\Gamma_u(t)=$$
$$=I\times\cup_{\iota=1}^r({}^1\Gamma_u\cap\partial\Omega^\iota), \tag{81}$$

$$\kappa_{ij}\frac{\partial T(\mathbf{x},t)}{\partial x_i}n_j=q\ (\text{or}=K(Y(\mathbf{x})-T(\mathbf{x},t)),\ u_i=\ {}^2u_{i2}\neq 0, \tag{82}$$
$$i,j=1,\dots,N,\ \text{on } {}^2\Gamma_u(t)=I\times\cup_{\iota=1}^r({}^2\Gamma_u\cap\partial\Omega^\iota),\ \ {}^2\Gamma_u=\cup_{i=2}^4\ {}^i\Gamma_u,$$

and the contact and the initial conditions

$$\left.\begin{array}{ll}
(a) & T^s=T^m,\\
 & \kappa_{ij}\frac{\partial T}{\partial x_i}n_{j|(s)}-\kappa_{ij}\frac{\partial T}{\partial x_i}n_{j|(m)}=\mathcal{F}_c^{sm}|\tau_n^{sm}|\,|[\mathbf{u}_t]^{sm}|,\\
(b) & [u_n]^{sm}\le d^{sm},\ \tau_n^s=\tau_n^m\equiv\tau_n^{sm}\le 0,\\
 & ([u_n]^{sm}-d^{sm})\tau_n^{sm}=0,\\
(c) & [\mathbf{u}_t]^{sm}=0\Longrightarrow|\boldsymbol{\tau}_t^{sm}|\le\mathcal{F}_c^{sm}(0)|\tau_n^{sm}|,\\
 & [\mathbf{u}_t]^{sm}\neq 0\Longrightarrow\boldsymbol{\tau}_t^{sm}=-\mathcal{F}_c^{sm}([\mathbf{u}_t]^{sm})|\tau_n^{sm}|\frac{[\mathbf{u}_t]^{sm}}{|[\mathbf{u}_t]^{sm}|},\\
 & \alpha=\alpha_0(\mathbf{x})+\alpha_D(\mathbf{x},\lambda^s,D),
\end{array}\right\}(t,\mathbf{x})\in I\times\cup_{s,m}\Gamma_c^{sm}, \tag{83}$$

$$T(\mathbf{x},0)=T_0(\mathbf{x}),\quad \mathbf{u}(\mathbf{x},0)=\mathbf{u}_0(\mathbf{x}),\quad \mathbf{x}\in\Omega, \tag{84}$$

where $\alpha=\alpha_0(\mathbf{x})\ge 0$ represents the elastic wave damping in the investigated region and $\alpha=\alpha_0(\mathbf{x})+\alpha_D(\mathbf{x},\lambda^s,D)$ at the moment when the frictional forces are not sufficient to preclude the mutual shifting of both geological bodies - the lithospheric plates or the lithospheric blocks in the subduction zone, and such that the propagated waves near the boundary Γ_u propagated through the distance D are fully damped, i.e., any reflected waves do not return back into the investigated domain, κ_{ij} are the non-linear heat conduction depending on the space variables, the temperature and the gradient of temperature, that are continuous in the sense of Carathéodory (i.e., $\mathbf{x} \mapsto \kappa_{ij}(\mathbf{x}, T, \nabla T)$ is measurable for every $T \in \mathbb{R}$ and $\mathbf{x} \mapsto \kappa_{ij}(\mathbf{x}, T, \nabla T)$ is continuous for almost every $\mathbf{x} \in \Omega$), locally Lipschitz with respect to ∇T, and satisfy

$$\begin{aligned} &\kappa_0(1+\delta_0 \mid T \mid^{\gamma_0} + \delta_1 \mid \nabla T \mid^{\gamma_1})\xi_i\xi_j \le \kappa_{ij}(\mathbf{x}, T, \nabla T) \le \\ &\le \kappa_1(1+\delta_0 \mid T \mid^{\gamma_0} + \delta_1 \mid \nabla T \mid^{\gamma_1})\xi_i\xi_j \\ &\text{for all } \{\xi_i\} \in \mathbb{R} \text{ and all } \mathbf{x} \in \Omega \text{ with } 0 < \kappa_0 \le \kappa_1 < +\infty, \\ &\gamma_0 > 0, \gamma_1 > 0, \delta_0 \ge 0, \delta_1 \ge 0, \end{aligned} \tag{85}$$

and the strong monotonicity condition and the continuity relation, that compensate the growth of the non-linear coupling terms (Eck et al. (2005); next we will assume that $\gamma_0 = \gamma_1 = \gamma \ge \frac{1}{2}$, $q = q(\mathbf{x}, t)$ is the heat flux, K the heat transition coefficient, $Y(\mathbf{x})$ the external temperature, $T_0(\mathbf{x}), T(\mathbf{x}, t)$ the initial and actual temperatures $\mathbf{u}(\mathbf{x}, t), \boldsymbol{\tau}(\mathbf{x}, t)$ are the velocity and the stress vectors, τ_{ij} the stress tensor, u_n, $\mathbf{u}_t$, $\tau_n, \boldsymbol{\tau}_t$ the normal and tangential components of velocity and stress vectors, $\mathbf{n}$ the outward normal to the boundary $\partial\Omega$ including Γ_c^{sm} and where $\mathbf{u}_0$, $\mathbf{u}_2$ are the given functions, where by $\mathbf{u}_2 = ({}^1\mathbf{u}_2, {}^2\mathbf{u}_2)$ we simulate the movements of the lithospheric plates (blocks) in the sense of the plate tectonics and the flows of the oceanic waters that are also connected with the typhoon, similarly as in the previous model problem. The coefficient of friction $\mathcal{F}_c^{sm} \equiv \mathcal{F}_c^{sm}(\mathbf{x}, \mathbf{u}_t)$ satisfies the similar conditions as in the previous section, and where $[\mathbf{u}]^{sm} = \mathbf{u}^s - \mathbf{u}^n$ is a jump (difference) of quantities $\mathbf{u}^s$ and $\mathbf{u}^m$ and d^{sm} is a gap, $d^{sm} \ge 0$, and represents the normalized gap between the contact boundaries of Ω^s and Ω^m (e.g. of both neighboring lithospheric plates or lithospheric blocks) or between the two faces of the fracture or deep fault, respectively, (i.e., Γ_c^s and Γ_c^m), and where on $\cup_{s,m}\Gamma_c^{sm}$ due to the equilibrium of forces $\tau_{ij}(\mathbf{u}^s, T^s)n_j^s = -\tau_{ij}(\mathbf{u}^m, T^m)n_j^m$ and the term $\mathcal{F}_c^{sm}|\tau_n^{sm}|\,|[\mathbf{u}_t]^{sm}|$ represents the frictional heatings on the contact surfaces Γ_c^{sm}. In the equation of heat conduction the second term on the right hand side represents the deformation heat and the third term on the right hand side represents the vicous heat. Furthermore, we denote by $(y)_+ \equiv \max\{y, 0\}$ the positive part of y.

To simplify the problem we can assume that the non-penetration conditions can be replaced by the penalty, having the physical meaning that the contact is not perfectly rigid and smooth. In addition, because our study is connected with the earthquake's mechanism, then in the approach based on the contact problem with visco-plastic friction in the thermo-visco-plastic rheology of the Bingham type the thermal part of the problem can be omitted. The origins of the earthquakes are arisen by (i) the mutual movements of neighboring lithospheric plates and blocks that being in contact or by (ii) the fracture mechanism. In the first case (i) the non-penetration conditions will be replaced by the bilateral contact type condition and the Coulomb law of friction in the form

$$[u_n]^{sm} = 0 \quad \text{and} \quad |\tau_t^{sm}| \leq \mathcal{F}_c^{sm}|\tau_n^{sm}|,$$

$$\left.\begin{cases} \text{if} \quad |\tau_t^{sm}| < \mathcal{F}_c^{sm}|\tau_n^{sm}| \quad \text{then} \quad [\mathbf{u}_t]^{sm} = 0, \\ \text{if} \quad |\tau_t^{sm}| = \mathcal{F}_c^{sm}|\tau_n^{sm}| \quad \text{then there exists } \lambda \geq 0 \\ \qquad \text{such that} \quad [\mathbf{u}_t]^{sm} = -\lambda\tau_t^{sm}, \\ \text{and } \alpha = \alpha_0(\mathbf{x}) + \alpha_D(\mathbf{x}, \lambda^s, D), \end{cases}\right\} \text{on } \cup_{s,m}\Gamma_c^{sm} \times I. \tag{86}$$

In the second case the earthquake's mechanism will be based on the modification of the fracture mechanics approach presented in the previous Subsection 3.5.

4.1.1. Variational (Weak) Solution of the Problem

We will introduce the Sobolev spaces of vector-functions having generalized derivatives of the (possibly fractional) order s of the type $[H^s(\Omega)]^k \equiv H^{s,k}(\Omega)$, where $H^{s,N}(\Omega) \equiv W_2^s(\Omega)$. The norm will be denoted by $\|\cdot\|_{s,k}$ and the scalar product by $(\cdot,\cdot)_s$ (for each integer k). We set $H^{0,k}(\Omega) \equiv L^{2,k}(\Omega)$. We introduce the space $C_0^\infty(\Omega)$ as a space of all functions in $C^\infty(\Omega)$ with a compact support in Ω^ι, $\iota = 1,\ldots,r$. The space is equipped with the ordinary countable system of semi-norms and as usual $C_0^\infty(\Omega,\mathbb{R}^N) = [C_0^\infty(\Omega)]^N$. We introduce the space $L^p(\Omega)$, $1 \leq p \leq \infty$, as the space of all measurable functions such that $\|f\|_{L^p(\Omega)} = (\int_\Omega |f(\mathbf{x})|^p d\mathbf{x})^{1/p} < +\infty$. By $L^\infty(\Omega)$ we denote the set of all measurable functions almost everywhere on Ω such that $\|f\|_\infty = \text{ess sup}_\Omega |f(\mathbf{x})|$ is finite. Let us introduce the Sobolev-Slobodenskii space $H_p^{s,N}(\Omega), 1 \leq p \leq \infty, s > 0, [s]$ its integer part, as in the previous section. Moreover, we define for $s \geq 1$, $s = \frac{1}{2}N$, $\gamma \geq \frac{1}{2}$, the following spaces and sets:

$$\begin{aligned}
{}^1H^{s,N}(\Omega) &= \prod_{\iota=1}^{r} {}^1H^{s,N}(\Omega^\iota) = \{\mathbf{v}|\mathbf{v} \in H^{s,N}(\Omega),\ \text{div}\,\mathbf{v} = 0 \text{ in } \cup_{\iota=1}^r \Omega^\iota, \\
&\qquad [v_n]^{sm} = 0 \text{ on } \cup_{s,m}\Gamma_c^{sm}\}, \\
{}^2H_{2+\gamma}^{s,1}(\Omega) &= \prod_{\iota=1}^{r} {}^2H_{2+\gamma}^{s,1}(\Omega^\iota) = \{z|z \in H_{2+\gamma}^{s,1}(\Omega), [z]^{sm} = 0 \text{ on } \cup_{s,m}\Gamma_c^{sm}\}, \\
{}^1V_s(\Omega) &= \{\mathbf{v}|\mathbf{v} \in {}^1H^{s,N}(\Omega), \mathbf{v}|_{\Gamma_u} = \mathbf{0}\}, \qquad {}^1V_s = \mathbf{u}_2 + {}^1V_s(\Omega), \\
{}^2V_s(\Omega) &= \{z|z \in {}^2H_{2+\gamma}^{s,1}(\Omega), z|_{\Gamma_\tau} = 0\}, \qquad {}^2V_s = T_1 + {}^2V_s(\Omega).
\end{aligned}$$

Then ${}^1H^{s,N}(\Omega)$ is a Hilbert space with the norm $\|\cdot\|_{s,N}$, ${}^2H_{2+\gamma}^{s,1}(\Omega)$ is a Hilbert space with the norm $\|\cdot\|_{s,1}$. For the sake of simplicity we put ${}^1H^{1,N}(\Omega) = {}^1H(\Omega)$, $\|\cdot\|_{1,N} \equiv \|\cdot\|_1$ and in $H^{0,N}(\Omega)$ $\|\cdot\|_{0,N} \equiv \|\cdot\|_0$. Let us put ${}^1\mathcal{H}(\Omega) = {}^1H^{1,N}(\Omega) \cap C_0^\infty(\Omega,\mathbb{R}^N)$, ${}^2\mathcal{H}(\Omega) = {}^2H_{2+\gamma}^{1,1}(\Omega) \cap C_0^\infty(\Omega)$. We will denote the dual space of ${}^1H^{s,N}(\Omega)$ by $({}^1H^{s,N}(\Omega))'$ and similarly in the second case. Furthermore, we put $\mathbf{v}' = \frac{\partial \mathbf{v}}{\partial t}$, $z' = \frac{\partial z}{\partial t}$,

$$\begin{aligned}
{}^1H &= \{\mathbf{v}|\mathbf{v} \in L^2(I;{}^1V_s), \mathbf{v}' \in L^2(I;\ {}^1H^{0,N}(\Omega)), \mathbf{v}(\mathbf{x},t_0) = 0\}, \\
{}^2H &= \{z|z \in L^{2+\gamma}(I;\ {}^2V_s), z' \in L^{2+\gamma}(I;\ {}^2H_{2+\gamma}^{0,1}(\Omega)), z(\mathbf{x},t_0) = T_0(\mathbf{x})\}.
\end{aligned}$$

Throughout this section we will assume that $\rho \in C(\overline{\Omega}), \rho \geq \rho_0 > 0, c_e \in C(\overline{\Omega}), c_e \geq e_{e0} > 0$, $\mathbf{f}(\mathbf{x},t) \in L^2(I;({}^1H^{1,N}(\Omega))')$, $\mathbf{P}(\mathbf{x},t) \in L^2(I;L^{2,N}(\Gamma_\tau))$, $\partial\beta_{ij}(\mathbf{x})/\partial x_j \in L^\infty(\Omega)$, $\forall i,j \in \{1,\ldots,N\}$, $q \in L^2(I;L^2({}^2\Gamma_1))$, $T_0^\iota(\mathbf{x}) \in {}^2H^{1,1}(\Omega^\iota)$, $\hat{g}$, $\hat{\mu}$ are piecewise constant and positive, $W \in L^2(I;({}^2H^{1,1}(\Omega))')$, κ_{ij} satisfy the above conditions, $Y \geq 0$, $Y \in L^2(\Gamma_u), K \geq 0$, $K \in L^\infty(\Gamma_u)$, $\alpha \in L^\infty(\overline{\Omega}), \alpha \geq \alpha_0 > 0$, $\mathcal{F}_c^{sm} \in L^\infty(\Gamma_c^{sm})$, $\mathcal{F}_c^{sm} \geq 0$ a.e. on $\cup_{s,m}\Gamma_c^{sm}$.

For $\mathbf{u},\mathbf{v} \in H^{1,N}(\Omega), T, z \in H^{1,1}_{2+\gamma}(\Omega)$ we put

$$(\mathbf{u}',\mathbf{v}) = \sum_{\iota=1}^r (\mathbf{u}^{\iota\prime},\mathbf{v}^\iota) = \int_\Omega \rho\mathbf{u}'\mathbf{v}d\mathbf{x}, \quad (\alpha\mathbf{u},\mathbf{v}) = \sum_{\iota=1}^r (\alpha\mathbf{u}^\iota,\mathbf{v}^\iota) = \int_\Omega \alpha\mathbf{u}\mathbf{v}d\mathbf{x},$$

$$a(\mathbf{u},\mathbf{v}) = \sum_{\iota=1}^r a^\iota(\mathbf{u}^\iota,\mathbf{v}^\iota) = 2\int_\Omega \hat{\mu}D_{ij}(\mathbf{u})D_{ij}(\mathbf{v})d\mathbf{x},$$

$$b(\mathbf{u},\mathbf{v},\mathbf{w}) = \sum_{\iota=1}^r b^\iota(\mathbf{u}^\iota,\mathbf{v}^\iota,\mathbf{w}^\iota) = \int_\Omega \rho u_i \frac{\partial v_j}{\partial x_i} w_j d\mathbf{x},$$

$$(T',z) = \sum_{\iota=1}^r (T^{\iota\prime},z^\iota) = \int_\Omega \rho c_e T' z d\mathbf{x},$$

$$a_T(w;T,z) = \sum_{\iota=1}^r a_T^\iota(w^\iota;T^\iota,z^\iota) = \int_\Omega \kappa_{ij}(w,\nabla w)\frac{\partial T}{\partial x_i}\frac{\partial z}{\partial x_j}d\mathbf{x},$$

$$S(\mathbf{v}) = \sum_{\iota=1}^r S^\iota(\mathbf{v}^\iota) = \int_\Omega f_i v_i d\mathbf{x} + \int_{\Gamma_\tau} P_i v_i ds \equiv (\mathbf{F},\mathbf{v}),$$

$$s(z) = \sum_{\iota=1}^r s^\iota(z^\iota) = \int_\Omega Wzd\mathbf{x} \ \ (\text{or} + \int_{{}^2\Gamma_u} qzds) \equiv (Q,z),$$

$$j(\mathbf{v}) = \sum_{\iota=1}^r j^\iota(\mathbf{v}^\iota) = 2\int_\Omega \hat{g}(D_{II}(\mathbf{v}))^{\frac{1}{2}}d\mathbf{x},$$

$$j_g(\mathbf{v}) = \sum_{\iota=1}^r j_g^\iota(\mathbf{v}^\iota) = \int_{\cup_{kl}\Gamma_c^{sm}} \mathcal{F}_c^{sm}|\tau_n^{sm}||\mathbf{v}_t^s - \mathbf{v}_t^m|ds,$$

$$b_K(T,z) = \langle K(T-Y),z\rangle_{{}^1\Gamma_u} = \sum_{\iota=1}^r \langle K^\iota(T^\iota - Y^\iota),z^\iota\rangle_{{}^1\Gamma_u\cap\partial\Omega^\iota}$$

$$= \int_{{}^1\Gamma_u} K(T-Y)zds,$$

$$j_T(z) = \sum_{\iota=1}^r j_T^\iota(z^\iota) = \langle \mathcal{F}_c^{sm} \mid \tau_n^{sm}(\mathbf{u},T) \mid\mid [\mathbf{u}_t]^{sm} \mid, z)\rangle_{\Gamma_c^{sm}} =$$

$$= \int_{\cup_{s,m}\Gamma_c^{sm}} \mathcal{F}_c^{sm}|\tau_n^{sm}(\mathbf{u},T)||[\mathbf{u}_t]^{sm}|zds,$$

$$b_s(T,\mathbf{v}) = \sum_{\iota=1}^r b_s^\iota(T^\iota,\mathbf{v}^\iota) = \int_\Omega \frac{\partial}{\partial x_j}(\beta_{ij}T)v_i d\mathbf{x},$$

$$b_p(\mathbf{v},z) = \sum_{\iota=1}^r b_p^\iota(\mathbf{v}^\iota,z^\iota) = \int_\Omega \rho T_0\beta_{ij}\frac{\partial v_i}{\partial x_j}zd\mathbf{x}.$$

Let us multiply (76) by $\mathbf{v}-\mathbf{u}(t)$ and (77) by $z-T(t)$, respectively, and further add both equations. We integrate the sum over Ω and apply the Green theorem satisfying the boundary conditions. Then after some modification, including among other the integration in time over the interval I, we obtain the following variational (weak) formulation:

Problem $(\mathcal{P})_v$: Find a pair of functions $(T,\mathbf{u})$ such that $T \in {}^2H$, $\mathbf{u} \in {}^1H$ and

$$\begin{aligned}
&\int_I [(\mathbf{u}'(t),\mathbf{v}-\mathbf{u}(t))+b(\mathbf{u}(t),\mathbf{u}(t),\mathbf{v}-\mathbf{u}(t))+(\alpha\mathbf{u}(t),\mathbf{v}-\mathbf{u}(t))+\\
&+(T'(t),z-T(t))+a(\mathbf{u}(t),\mathbf{v}-\mathbf{u}(t))+a_T(T(t);T(t),z-T(t))+\\
&+b_s(T(t)-T_0,\mathbf{v}-\mathbf{u}(t))+b_p(\mathbf{u}(t),z-T(t))+b_K(T(t),z-T(t))+\\
&+j(\mathbf{v})-j(\mathbf{u}(t))+j_g(\mathbf{v})-j_g(\mathbf{u}(t))+j_T(z)-j_T(T(t))]dt \geq\\
&\geq \int_I [S(\mathbf{v}-\mathbf{u}(t))+s(z-T(t))]dt \quad \forall(\mathbf{v},z) \in {}^1H \times {}^2H\\
&\text{holds for a.a. } t \in I. \qquad (87)
\end{aligned}$$

Using the above assumptions and

$$\text{since } \mathcal{F}_c^{kl} \in L^\infty(\Gamma_c^{kl}), \quad \mathcal{F}_c^{kl} \geq 0, \quad \text{then} \quad j_g(\mathbf{u}(t)) \geq 0\,, j_T(z(t)) \geq 0\,,$$

then the main result is represented by the following theorem:

Theorem 2. *Let $N \geq 2$, $s=\frac{1}{2}N$, $\gamma \geq \frac{1}{2}$. Under the above assumptions about the physical quantities, then there exists a pair of functions $(\mathbf{u},T)$ such that*

$$\begin{aligned}
&\mathbf{u} \in L^2(I;\, {}^1V_s) \cap L^\infty(I;\, {}^1V_0), \mathbf{u}' \in L^2(I;\, {}^1V_s)')\,,\\
&T \in L^{2+\gamma}(I;\, {}^2V_s) \cap L^\infty(I;\, {}^2V_0), T' \in L^{2+\gamma}(I;\, {}^2V_s')\,,\\
&\mathbf{u}(\mathbf{x},t_0)=0, \quad T(\mathbf{x},t_0)=T_0(\mathbf{x})
\end{aligned}$$

and satisfying the variational inequality (87).

Proof: To prove the visco-plastic part of the theorem the multiple regulatizations will be used. Then the proof is similar to that of Duvaut and Lions (1976), Nedoma (1995a,b), (1998), (1999), (2003b) (2006a), (2010), (2012a). Uniqueness of the problem for $N=3$ is an open problem. To prove the thermal part of the problem then modifications of the approaches of Eck et al. (2005) can be used.

Although the Bingham rheology deals with fluids, it is also called the "Bingham solid" (see Oldroyd (1947)), as the Bingham rheology lies in its liquid/solid character, that is, in the existence of the rigid zones located in the interior of the flows. Since the yield limit $\hat{g}$ increases, the rigid zones become larger and larger and thus it can completely block the flow (Duvaut and Lions (1976), Glowinski et al. (1981)). The solid rocks are blocked in their natural configurations and at the moment of the beginning of a flow, the lithospheric plate starts to move as the friction decreases or the slope starts to slide down. Therefore, the certain safety factor for obtaining a quantitative and qualitative evaluation of the blocking phenomenon was introduced by Hassani et al. (2005a,b) (Proposition 4.2). Hild et al. (2002) study the link between the yield limit distribution and the external forces and mass density distributions for which the flow of the Bingham fluid is blocked. If the mass density is assumed to be a function of time, i.e., $\rho = \rho(\mathbf{x},t)$, in these type of problems we find the displacement velocity field and the mass density that is defined by the conservation of mass. As a result they introduce, for the case when the volume forces are independent of time, that is, $\mathbf{f} = \mathbf{f}(\mathbf{x})$, and under the assumption that the density has no time evolution, that is, $\rho = \rho(\mathbf{x})$, the following statements:

Proposition 1. *For the blocking property the following three statements are equivalent:*
(i) The Bingham fluid is blocked if $\mathbf{u} = \mathbf{0}$, *where* $\mathbf{u}$ *is the displacement velocity field.*
(ii) The blocking inequality

$$\int_\Omega \hat{g}(\rho_0(\mathbf{x}))\|D\mathbf{v}\|d\mathbf{x} = \int_\Omega \rho_0(\mathbf{x})\mathbf{v}(\mathbf{x})d\mathbf{x} \qquad \forall \mathbf{v} \in {}^1V$$

holds.
(iii) There exists $\boldsymbol{\tau} \in \mathcal{H}_\sigma^b(\Omega)$ *such that* $|\boldsymbol{\tau}(\mathbf{x})| \leq \hat{g}(\mathbf{x})$ *a.e.* $\mathbf{x} \in \Omega$,
where
$\mathcal{H}_\sigma^b(\Omega) = \{\boldsymbol{\sigma} \in \mathcal{H}_\sigma(\Omega);\ \exists p \in L^2(\Omega),\ div\,\boldsymbol{\sigma} - \nabla p = -\mathbf{f}\ in\ \Omega,\ (\boldsymbol{\sigma} - p\mathbf{I}).\mathbf{n} = \mathbf{P}\ on\ \Gamma_\tau\}$,
$\mathcal{H}_\sigma(\Omega) = \{\boldsymbol{\sigma} \in L^{2,N\times N}(\Omega);\ \sigma_{ij} = \sigma_{ji};\ trace(\boldsymbol{\sigma}) = 0\ \in \Omega\}$.

4.1.2. Numerical Approach

The investigated variational problem $(\mathcal{P})_v$ will be solved numerically in its dynamical and/or stationary (i.e., when the mass movements are uniform) flow formulations. Next, the ideas of numerical solution of dynamic flow formulation will be discussed only.

Then for the thermal part of the model we apply the secant modulus method similarly as in the previous section. Then we have to solve the following problem:

$$\begin{aligned}
&(T'_{m+1}, z - T_{m+1}(t)) + a^{(T)}(T_m; T_{m+1}(t), z - T_{m+1}(t)) + \\
&+ b_p(\mathbf{u}_{m+1}(t), z - T_{m+1}(t)) \geq \\
&\geq (W, z - T_{m+1}(t)) + \langle K(T_{m+1} - Y), z - T_{m+1}(t)\rangle_{\,{}^1\Gamma_u} + \\
&+ \langle \mathcal{F}_c^{sm} \mid \tau_n^{sm} \mid\mid \mathbf{u}_{t,m} \mid, z - T_{m+1}(t)\rangle_{\Gamma_c^{sm}} \quad \forall z \in {}^2V\,.
\end{aligned}$$

Next, for simplicity we put $T_{m+1} \equiv T,\ T_m \equiv \overline{T},\ \mathbf{u}_{m+1} \equiv \mathbf{u},\ \mathbf{u}_m \equiv \overline{\mathbf{u}}$.

Let Ω_h be a polyhedral approximation to Ω in $\mathbb{R}^3$ and let its boundary be denoted as $\partial\Omega_h = \Gamma_{uh} \cup \Gamma_{\tau h} \cup \Gamma_{ch}$. Let $\mathfrak{T}_h$ be a partition of $\overline{\Omega}_h$ by tetrahedra $\mathcal{T}_{hi}$. Let $h = h(\mathcal{T}_{hi})$ be the maximum diameter of tetrahedral elements $\mathcal{T}_{hi}$. Let $\mathcal{T}_{hi} \in \mathfrak{T}_h$ be a tetrahedron with vertices P_i, $i = 1,\ldots,4$ and let R_i be the barycentres with respect to the points P_i, $i = 1,\ldots,4$. Assume that $\{\mathfrak{T}_h\}$ is a regular family of partitioning $\mathfrak{T}_h$ of Ω_h such that $\overline{\Omega}_h = \cup_{\mathcal{T}_{hi} \in \mathfrak{T}_h} \mathcal{T}_h$.

As regards the spatial discretization, there are several classes of possibilities for approximation ${}^1V_s$ by its finite element space ${}^1V_h$ (Glowinski et al. (1976), Temam (1977), (1979), Nedoma (1998), (2006a)). In this subsection spaces ${}^1V_s$, ${}^2V_s$ are approximated by spaces of linear non-conforming functions ${}^1V_h$ in the visco-plastic part of the problem and by spaces of linear conforming functions ${}^2V_h$ in the thermal part of the problem. Let ${}^1\mathcal{V}_h = \{\mathbf{v}_h | v_{hi} \in P_1^*,\ i = 1,\ldots,N,\ \forall \mathcal{T}_{hi} \in \mathfrak{T}_h$, continuous in barycentres of tetrahedra B_j, $j = 1,\ldots,m$, and equal to zero in B_j laying on Γ_{uh}, $\sum_{i=1,\ldots,N} \partial v_{hi}/\partial x_i = 0$,

$\mathbf{v}_h = 0,\ \mathbf{x} \in \Omega\backslash\overline{\Omega}_h\}$, where P_1^* denotes the space of all non-conforming linear polynomials, ${}^1V_h = \{\mathbf{v}_h | \mathbf{v}_h \in {}^1\mathcal{V}_h,\ \mathbf{v}_h = \mathbf{u}_1 \text{ on } \Gamma_{uh}\}$ and similarly, let ${}^2\mathcal{V}_h$, ${}^2V_h$ be the spaces of conforming linear finite elements, defined by the usual way as in the previous section.

For $\mathbf{u}_h, \mathbf{v}_h \in H^{1,N}(\Omega_h), T_h, z_h \in H^{1,1}_{2+\gamma}(\Omega_h)$ we put

$$(\mathbf{u}'_h, \mathbf{v}_h) = \sum_{\iota=1}^{r} (\mathbf{u}_h^{\iota\prime}, \mathbf{v}_h^{\iota}) = \int_{\Omega_h} \rho_h \mathbf{u}'_h \mathbf{v}_h d\mathbf{x},$$

$$(\alpha_h \mathbf{u}_h, \mathbf{v}_h) = \sum_{\iota=1}^{r} (\alpha_h \mathbf{u}_h^{\iota}, \mathbf{v}_h^{\iota}) = \int_{\Omega_h} \alpha_h \mathbf{u}_h \mathbf{v}_h d\mathbf{x},$$

$$a_h(\mathbf{u}_h, \mathbf{v}_h) = \sum_{\iota=1}^{r} a_h^{\iota}(\mathbf{u}_h^{\iota}, \mathbf{v}_h^{\iota}) = 2\int_{\Omega_h} \hat{\mu}_h D_{ij}(\mathbf{u}_h) D_{ij}(\mathbf{v}_h) d\mathbf{x},$$

$$b_h(\mathbf{u}_h, \mathbf{v}_h, \mathbf{w}_h) = \sum_{\iota=1}^{r} b_h^{\iota}(\mathbf{u}_h^{\iota}, \mathbf{v}_h^{\iota}, \mathbf{w}_h^{\iota}) = \int_{\Omega} \rho_h u_{hi} \frac{\partial v_{hj}}{\partial x_i} w_{hj} d\mathbf{x},$$

$$(T'_h, z_h) = \sum_{\iota=1}^{r} (T_h^{\iota\prime}, z_h^{\iota}) = \int_{\Omega_h} \rho_h c_{eh} T'_h z_h d\mathbf{x},$$

$$a_{Th}(w_h; T_h, z_h) = \sum_{\iota=1}^{r} a_{Th}^{\iota}(w_h^{\iota}; T_h^{\iota}, z_h^{\iota}) = \int_{\Omega_h} \kappa_{hij}(w_h, \nabla w_h) \frac{\partial T_h}{\partial x_i} \frac{\partial z_h}{\partial x_j} d\mathbf{x},$$

$$S_h(\mathbf{v}_h) = \sum_{\iota=1}^{r} S^{\iota}(\mathbf{v}_h^{\iota}) = \int_{\Omega_h} f_{hi} v_{hi} d\mathbf{x} + \int_{\Gamma_{\tau h}} P_{hi} v_{hi} ds \equiv (\mathbf{F}_h, \mathbf{v}_h),$$

$$s_h(z_h) = \sum_{\iota=1}^{r} s^{\iota}(z_h^{\iota}) = \int_{\Omega_h} W_h z_h d\mathbf{x}\ (\text{or} + \int_{{}^1\Gamma_{uh}} q_h z_h ds) \equiv (Q_h, z_h),$$

$$j_h(\mathbf{v}_h) = \sum_{\iota=1}^{r} j_h^{\iota}(\mathbf{v}_h^{\iota}) = 2\int_{\Omega_h} \hat{g}_h (D_{II}(\mathbf{v}_h))^{\frac{1}{2}} d\mathbf{x},$$

$$j_{gh}(\mathbf{v}_h) = \sum_{\iota=1}^{r} j_{gh}^{\iota}(\mathbf{v}_h^{\iota}) = \int_{\cup_{sm}\Gamma_{ch}^{sm}} \mathcal{F}_{ch}^{sm} |\tau_n^{sm}| \, |[\mathbf{v}_{th}]^{sm}| ds,$$

$$j_{Th}(z_h) = \sum_{\iota=1}^{r} j_{Th}^{\iota}(z_h^{\iota}) = \langle \mathcal{F}_{ch}^{sm} \mid \tau_n^{sm}(\overline{\mathbf{u}}_h, T_h) \mid\mid [\overline{\mathbf{u}}_{ht}]^{sm} \mid, z_h) \rangle_{\Gamma_{ch}^{sm}} =$$

$$= \int_{\cup_{s,m}\Gamma_{ch}^{sm}} \mathcal{F}_{ch}^{sm} |\tau_n^{sm}(\overline{\mathbf{u}}_h, T_h)| \, |[\overline{\mathbf{u}}_{ht}]^{sm}| z_h ds,$$

$$b_{Kh}(T_h, z_h) = \langle K_h(T_h - Y_h), z_h \rangle_{{}^1\Gamma_{uh}} =$$

$$= \sum_{\iota=1}^{r} \langle K_h^{\iota}(T_h^{\iota} - Y_h^{\iota}), z_h^{\iota} \rangle_{{}^1\Gamma_{uh} \cap \partial\Omega_h^{\iota}} = \int_{{}^1\Gamma_{uh}} K_h(T_h - Y_h) z_h ds,$$

$$b_{sh}(T_h, \mathbf{v}_h) = \sum_{\iota=1}^{r} b_{sh}^{\iota}(T_h^{\iota}, \mathbf{v}_h^{\iota}) = \int_{\Omega_h} \frac{\partial}{\partial x_j} (\beta_{hij} T_h) v_{hi} d\mathbf{x},$$

$$b_{ph}(\mathbf{v}_h, z_h) = \sum_{\iota=1}^{r} b_{ph}^{\iota}(\mathbf{v}_h^{\iota}, z_h^{\iota}) = \int_{\Omega_h} \rho_h T_{0h} \beta_{hij} \frac{\partial v_{hi}}{\partial x_j} z_h d\mathbf{x}.$$

Then we have the following discrete problem:

Problem$(\mathcal{P}_h)$: Find a pair of functions $(\mathbf{u}_h, T_h)$ such that $\mathbf{u}_h \in {}^1V_h$, $T_h \in {}^2V_h$ and

$$\begin{aligned}
&(\mathbf{u}_h'(t), \mathbf{v}_h - \mathbf{u}_h(t)) + b_h(\mathbf{u}_h(t), \mathbf{u}_h(t), \mathbf{v}_h) + (\alpha \mathbf{u}_h(t), \mathbf{v}_h - \mathbf{u}_h(t)) + \\
&+ a_h(\mathbf{u}_h(t), \mathbf{v}_h - \mathbf{u}_h(t)) + b_{sh}(T_h(t) - T_0, \mathbf{v}_h - \mathbf{u}_h(t) + \\
&+ j_h(\mathbf{v}_h(t)) - j_h(\mathbf{u}_h(t)) + j_{gh}(\mathbf{v}_h(t)) - j_{gh}(\mathbf{u}_h(t)) \geq \\
&\geq S(\mathbf{v}_h - \mathbf{u}_h(t)) \quad \forall (\mathbf{v}_h, z_h) \in {}^1V_h \qquad (88)
\end{aligned}$$

holds for a.a. $t \in I$, (because $b(\mathbf{v}, \mathbf{v}, \mathbf{v}) = 0$),

$$\begin{aligned}
&(T_h'(t), z_h - T_h(t)) + a_{Th}(\overline{T}_h; T_h(t), z_h - T_h(t)) + \\
&+ b_{ph}(\mathbf{u}_h(t), z_h - T_h(t)) + b_{Kh}(T_h(t), z_h - T_h(t)) + \\
&+ j_{Th}(z(t)) - j_{Th}(T_h(t)) \geq s_h(z_h - T_h(t)) \quad \forall z_h \in {}^2V_h \qquad (89)
\end{aligned}$$

holds for a.a. $t \in I$.

Now, let us introduce the main idea how to solve the problem $(\mathcal{P}_h)$, that will be modification of that of Nedoma (2006a), (2010), (2012a), Nedoma and Tomášek (2008). Let n be an integer and set $k = t_p/n$. Let

$$\mathbf{f}_k^{i+\Theta} = \frac{1}{k}\int_{ik}^{(i+\Theta)k} \mathbf{f}(t)dt, \quad i = 0, \ldots, n-1, \quad 0 \leq \Theta \leq 1, \quad \mathbf{f}_k^{i+\Theta} \in {}^1V_s',$$

$$W_k^{i+\Theta} = \frac{1}{k}\int_{ik}^{(i+\Theta)k} W(t)dt, \quad i = 0, \ldots, n-1, \quad 0 \leq \Theta \leq 1, \quad W_k^{i+\Theta} \in {}^2V_s'.$$

We will start with the initial data $\mathbf{u}^0 = \mathbf{u}_0$ ($e.g.$ $= \mathbf{0}$; or $\neq \mathbf{0}$ - in this case determining the distributions of the stress-strain rate fields from the last earthquake in the regions of our interest, the subduction zone and the closed regions), $T^0 = T_0$. When $(\mathbf{u}^0, T^0), \ldots, (\mathbf{u}^i, T^i)$ are known, we define $(\mathbf{u}^{i+1}, T^{i+1})$ as an element of ${}^1V_s \times {}^2V_s$. The existence of a pair of functions $(\mathbf{u}^{i+1}, T^{i+1})$ for each fixed k and each $i \geq 0$ can be proved.

Given a triangulation $\mathcal{T}_h$ of Ω_h, we assume $(\mathbf{u}_h^0, T_h^0)$ to be given in ${}^1V_h \times {}^2V_h$, and, taken by such a way that $(\mathbf{u}_h^0, T_h^0)$ tends to $(\mathbf{u}_{0h}, T_{0h})$ in ${}^1V_h \times {}^2V_h$ when $h \to 0_+$. Moreover, we will assume that $\|\mathbf{u}_h^0\|$, $\|T_h^0\|$ are bounded. We define recursively for each k and h a family of elements $(\mathbf{u}_h^0, T_h^0), \ldots, (\mathbf{u}_h^i, T_h^i)$ of ${}^1V_h \times {}^2V_h$, which can be based on implicit, semi-implicit and/or explicit schemes. In our paper the semi-implicit scheme will be introduced and shortly discussed. The time derivatives will be approximated by the backward differences. Since $\frac{\mathbf{u}_h^{i+1} - \mathbf{u}_h^i}{k} = \frac{\mathbf{u}_h^{i+\Theta} - \mathbf{u}_h^i}{\Theta k}$, $\frac{T_h^{i+1} - T_h^i}{k} = \frac{T_h^{i+\Theta} - T_h^i}{\Theta k}$, $0 \leq \Theta \leq 1$, then

$$\mathbf{u}_h^{i+\Theta} = \Theta \mathbf{u}_h^{i+1} + (1-\Theta)\mathbf{u}_h^i, \quad T_h^{i+\Theta} = \Theta T_h^{i+1} + (1-\Theta) T_h^i, \quad 0 \leq \Theta \leq 1,$$

$$\mathbf{u}_h^{i+1} = \Theta^{-1}\mathbf{u}_h^{i+\Theta} - \frac{(1-\Theta)}{\Theta}\mathbf{u}_h^i, \quad T_h^{i+1} = \Theta_h^{-1} T_h^{i+\Theta} - \frac{(1-\Theta)}{\Theta} T_h^i, \quad 0 \leq \Theta \leq 1.$$

The contact conditions depend on the solution of the studied problem, therefore, it can be assumed that $j_{gh}(\mathbf{v}_h) = j_{gh}(\mathbf{u}_h^i, \mathbf{v}_h)$.

Scheme $(\mathcal{P}_{si})$: When $(\mathbf{u}_h^0, T_h^0), \ldots, (\mathbf{u}_h^i, T_h^i)$ are known, then determine $(\mathbf{u}_h^{i+1}, T_h^{i+1})$ as a solution in ${}^1V_h \times {}^2V_h$ by the following scheme

$$\begin{aligned}&k^{-1}(\mathbf{u}_h^{i+1} - \mathbf{u}_h^i, \mathbf{v}_h - \mathbf{u}_h^{i+\Theta}) + (\alpha_h \mathbf{u}_h^{i+\Theta}, \mathbf{v}_h - \mathbf{u}_h^{i+\Theta}) +\\ &b_h(\mathbf{u}_h^{i+\Theta}, \mathbf{u}_h^{i+\Theta}, \mathbf{v}_h) + a_h(\mathbf{u}_h^{i+\Theta}, \mathbf{v}_h - \mathbf{u}_h^{i+\Theta}) + b_s(T_h^i - T_h^0, \mathbf{v}_h - \mathbf{u}_h^{i+\Theta}) +\\ &+ j_h(\mathbf{v}_h) - - j_h(\mathbf{u}_h^{i+\Theta}) + j_{gh}(\mathbf{v}_h) - j_{gh}(\mathbf{u}_h^{i+\Theta}) \geq\\ &\geq (\mathbf{F}_h^{i+\Theta}, \mathbf{v}_h - \mathbf{u}_h^{i+\Theta}) \quad \forall \mathbf{v}_h \in {}^1V_h, \end{aligned} \tag{90}$$

$$\begin{aligned}&k^{-1}(T_h^{i+1} - T_h^i, z_h - T_h^{i+\Theta}) + a_{Th}(\overline{T}_h; T_h^{i+\Theta}, z_h - T_h^{i+\Theta}) + j_T(z_h) -\\ &- j_T(T_h^{i+\Theta})_+ b_p(\overline{\mathbf{u}}_h, z_h - T_h^{i+\Theta}) + b_K(T_h^{i+\Theta}, z_h - T_h^{i+\Theta}) \geq\\ &\geq (Q_h^{i+\Theta}, z_h - T_h^{i+\Theta}) \quad \forall z_h \in {}^2V_h. \end{aligned} \tag{91}$$

The above scheme is valid for $\Theta \geq \frac{1}{2}$; for $\Theta = 1$ we have the semi-implicit scheme and for $\Theta = \frac{1}{2}$ we have the Crank-Nicolson scheme. The other schemes that can be applied are discussed in Glowinski (2003). Note that $b(\mathbf{v}, \mathbf{v}, \mathbf{v}) = 0$ as usual (see Temam (1977)).

Because the term $\hat{g} D_{ij}(\mathbf{u}) D_{II}^{-\frac{1}{2}}(\mathbf{u})$ in the variational formulation represents the non-differential functional $j(\mathbf{u})$ and because $\|j_\varepsilon''(\mathbf{u})\|_* \simeq \frac{1}{\varepsilon}$ (see Glowinski (2003)), then the Problem $(\mathcal{P})_v$ may be badly conditioned. Therefore, we will introduce the multipliers $m = (m_{ij})$, $i, j = 1 \dots N$, defined by (see Duvaut and Lions (1976), Chapter 6, Section 9 and Glowinski (2003), Section 50)

$$\begin{aligned}&m_{ij} \in L^\infty(\Omega \times I), \ m_{ij} = m_{ji} \ \forall i, j, \ m_{ii} = 0,\\ &m_{ij} m_{ij} \leq 1 \ \text{a.e. in } \Omega \times I,\\ &m_{ij} D_{ij}(\mathbf{u}) = (D_{ij}(\mathbf{u}) D_{ij}(\mathbf{u}))^{\frac{1}{2}} = 2^{\frac{1}{2}} D_{II}^{\frac{1}{2}}(\mathbf{u}) \ \text{a.e. in } \Omega \times I.\end{aligned}$$

Then the multipliers $m \in K = \{m | m_{ij} \in L^\infty(\Omega \times I), \|m\|_\infty \leq 1\}$ can be approximated by tensor's functions, that are piecewise constant on every $\mathcal{T}_{hi} \in \mathcal{T}_h$. Moreover, the space K will be approximated by

$$\begin{aligned}K_h = \{&m_h | \, \|m_h\|_\infty \leq 1, \ m_{hjl} \text{ piecewise constant on every } \mathcal{T}_{hi},\\ &m_{hjl} = 0 \text{ on } \Omega \backslash \Omega_h, \ j, l = 1 \dots N, \ m_{hjl} = m_{hlj}\}.\end{aligned}$$

Resulting of all previous results we obtain the following scheme:

Problem $(\mathcal{P})_{\mathrm{m}}$: Let us assume that $\mathbf{u}_h^0, \mathbf{u}_h^1 \dots \mathbf{u}_h^i$ be known. Then we determine $\mathbf{u}_h^{i+1}, \mathbf{u}_h^{i+\Theta}, m_h^{i+\Theta}$ as follows

$$\begin{aligned}&m_{hjl}^{i+\Theta} m_{hjl}^{i+\Theta} \leq 1 \text{ on every } \mathcal{T}_{hi} \in \mathfrak{T}_h,\\ &m_{hjl}^{i+\Theta} = m_{hlj}^{i+\Theta}, \ m_{hjl}^{i+\Theta} \in L^\infty(\Omega), \ 1 \leq j, \ l \leq N,\\ &m_{hjl}^{i+\Theta} D_{jl}(\mathbf{u}_h^{i+\Theta}) = \left(D_{jl}(\mathbf{u}_h^{i+\Theta}) D_{jl}(\mathbf{u}_h^{i+\Theta})\right)^{\frac{1}{2}} \text{ on every } \mathcal{T}_{hi} \in \mathfrak{T}_h,\end{aligned}$$

$$\begin{aligned}&(\mathbf{u}_h^{i+\Theta}, \mathbf{v}_h) + (\alpha_h \mathbf{u}_h^{i+\Theta}, \mathbf{v}_h) + k\Theta a_h(\mathbf{u}_h^{i+\Theta}, \mathbf{v}_h) + k\Theta b_h(\mathbf{u}_h^i, \mathbf{u}_h^i, \mathbf{v}_h) +\\ &+ 2^{\frac{1}{2}} k\Theta \hat{g}\left(m_{hjl}^{i+\Theta}, D_{jl}(\mathbf{v}_h)\right) + k\Theta j_{gh}(\mathbf{v}_h) = k\Theta(\mathbf{F}_h^{i+\Theta}, \mathbf{v}_h) + (\mathbf{u}_h^i, \mathbf{v}_h)\\ &\forall \mathbf{v}_h \in {}^1V_h, \ i = 0, 1 \dots n-1,\end{aligned} \tag{92}$$

and where $\mathbf{u}_h^{i+1}$ will be determined from $\mathbf{u}_h^{i+1} = \Theta^{-1} \mathbf{u}_h^{i+\Theta} + {}+\Theta^{-1}(1 - \Theta)\mathbf{u}_h^i$.

Let us denote by $|\cdot|_h$ the norm in $L^2(\Omega)$ (or $[\mathrm{L}^2(\Omega)]^{\mathrm{N}}$), by $\|\cdot\|_h$ the norms in ${}^1\mathcal{V}_h$, ${}^1V_h$ and ${}^2\mathcal{V}_h$, ${}^2V_h$. Using the similar techniques as in Temam (1977), (1979), Nedoma (2006a) $|\mathbf{u}_h|_h \le d_1\|\mathbf{u}_h\|_h$, $\|\mathbf{u}_h\|_h \le S(h)|\mathbf{u}_h|_h$ $\forall \mathbf{u}_h \in {}^1\mathcal{V}_h$, $|T_h|_h \le d_0\|T_h\|_h$, $\|T_h\|_h \le S_0(h)$ $|T_h|_h$ $\forall T_h \in {}^2\mathcal{V}_h$, where d_0, d_1 are independent of h, and moreover, $|b_h(\mathbf{u}_h,\mathbf{u}_h,\mathbf{v}_h)| \le S_1(h)|\mathbf{u}_h|_h\|\mathbf{u}_h\|_h|\mathbf{v}_h|_h$ $\forall \mathbf{u}_h,\mathbf{v}_h \in {}^1V_h$, where $S_1(h) \le d_1 S_1^2(h)$ and $b_h(\mathbf{u}_h,\mathbf{u}_h,\mathbf{u}_h) = 0$ $\forall \mathbf{u}_h \in {}^1V_h$, and

$$|b_{sh}(T_h - T_0,\mathbf{u}_h) + b_{ph}(\mathbf{u}_h,T_h)| \le c(1+\|T_h\|_h|\mathbf{u}_h|_h + |T_h|_h\|\mathbf{u}_h\|_h)$$
$$\forall T_h \in {}^2\mathcal{V}_h,\ \forall \mathbf{u}_h \in {}^1\mathcal{V}_h,$$

Moreover, we have

$$|(m_h, D\mathbf{v}_h)| \le |m_h|\,|D\mathbf{v}_h| = \left(\int_\Omega \sum_{j,l=1}^N (m_{hjl})^2 d\mathbf{x}\right)^{\frac{1}{2}} \left(\int_\Omega \sum_{j,l=1}^N (D_{jl}(\mathbf{v}_h))^2 d\mathbf{x}\right)^{\frac{1}{2}}.$$

Because

$$\int_\Omega \sum_{j,l=1}^N (D_{jl}(\mathbf{v}_h))^2 d\mathbf{x} \le \|\mathbf{v}_h\|_{1,N}^2,\quad \int_\Omega \sum_{j,l=1}^N (m_{hjl})^2 d\mathbf{x} \le \mathrm{meas}\,(\Omega)\ \forall m_h \in K_h,$$

then we have

$$\left|k\Theta 2^{\frac{1}{2}}\hat{g}\,(m_h, D\mathbf{v}_h)\right| \le \left(k\Theta 2^{\frac{1}{2}}\hat{g}\,(\mathrm{meas}\,(\Omega))\right)^{\frac{1}{2}} \|\mathbf{v}_h\|_{1,N}^2 \quad \forall m_h \in K_h, \forall \mathbf{v}_h \in {}^1V_h.$$

Summarize all of the obtained results it can be shown that $(\mathbf{u}_h^i, T_h^i)$ and $(\mathbf{u}_h^{i+1}, T_h^{i+1})$ defined by the Scheme $(\mathcal{P}_{si})$, satisfy for $\Theta \ge \frac{1}{2}$ the following conditions:

Proposition 2. *Let the family of triangulations $\{\mathfrak{T}_h\}$ be uniformly regular, and let the angles in the tetrahedra be less or equal to $\frac{\pi}{2}$. Let k, h satisfy $kS_0(h) \le d_0$, $kS(h) \le d_1$, where d_0, d_1 are positive constants independent of k, h. Let $\Theta \ge \frac{1}{2}$. Then $(T_h^i, \mathbf{u}_h^i)$ are defined by the scheme $(\mathcal{P}_{si})$ and*

$$|\mathbf{u}_h^i|^2 \le c, \quad i = 0,\dots,n, \quad k\sum_{i=0}^{n-1} \|\mathbf{u}_h^{i+\Theta}\|^2 \le c,$$
$$\sum_{i=0}^{n-1} |\mathbf{u}_h^{i+1} - \mathbf{u}_h^i|^2 \le c, \quad k\sum_{i=0}^{n-1} \left\|\mathbf{u}_h^{i+1} - \mathbf{u}_h^i\right\|^2 \le c, \tag{93}$$

$$|T_h^i|^2 \le c_0, \quad i = 0,\dots,n, \quad k\sum_{i=0}^{n-1} \|T_h^{i+\Theta}\|^2 \le c_0,$$
$$\sum_{i=0}^{n-1} |T_h^{i+1} - T_h^i|^2 \le c, \quad k\sum_{i=0}^{n-1} \|T_h^{i+1} - T_h^i\|^2 \le c_0, \tag{94}$$

hold, where c, c_0 are constants independent of k, h.

The scheme $(\mathcal{P}_{si})$ is stable and convergent, and moreover, in addition due to the convergence of the secant modulus method the scheme of the thermal part of the problem is also convergent. The proofs of the individual parts of the problem are similar of that of Glowinski et al. (1976), Nedoma (1998), Han et al. (1997) and Falleta and Lamichhane (2009). The difficulty in practical computations is connected with the approximation of the constraint div $\mathbf{v}_h = 0$, i.e., the incompressibility condition of the Bingham's fluid. For its approximation see e.g. Glowinski et al. (1976), Temam (1979), Nedoma (1998).

Remark 9. *The studied problem can be also solved by using the penalty and the regularization techniques. When we assume that the mass movements are uniform then the problem corresponds to the quasi-stationary flow of the thermo-Bingham fluid.*

Algorithm

We saw above that the problem is nonlinear due to the term $j_h(\mathbf{v}_h)$ (in Nedoma (1998), (2006a), Nedoma and Tomášek (2008)), and moreover, the functionals $j_h(\mathbf{v}_h)$ and $j_{gh}(\mathbf{v}_h)$ are non-differentiable functionals. For simplicity we will assume similarly as in the previous section that $j_{gh}(\mathbf{v}_h)$ will depend on velocities in the previous time level(s). The great difficulties in practical computations are connected with the approximation of the incompressibility constraint $\operatorname{div}\mathbf{v}_h = 0$ and with the discrete approximation of the pressure. In the first case it lies in finding the basis of the space V_h with $\operatorname{div}\mathbf{v}_h = 0$. For these we can use the Uzawa or the Arrow-Hurwicz algorithms (Temam (1977), (1979), Nedoma (1998)). Because the functional $j_h(\mathbf{v_h})$ is a non-differential functional, then the problem leads to finding a saddle point, therefore the Uzawa algorithm will be also used to finding a saddle point of the Lagrangian introduced below (Temam (1977)). Thus, in every step of Uzawa's algorithm the system of linear algebraic equations with sparse positive definite matrix is solved by using the conjugate gradient method or the preconditioned conjugate gradient methods can be also used. The Uzawa's algorithms can be compiled into one algorithm only. For more details see Hladik and Nedoma (1992) and Nedoma (1998, Chapter 8).

Let us introduce the bilinear form $A_h(\mathbf{v}_h, \mathbf{v}_h)$ and the operator $\overline{A}_h$ on ${}^1\mathcal{V}_\mathrm{h}$ by

$$A_h(\mathbf{u}_h, \mathbf{v}_h) = (\mathbf{u}_h, \mathbf{v}_h) + k\Theta a_h(\mathbf{u}_h, \mathbf{v}_h) \quad \forall \mathbf{u}_h, \mathbf{v}_h \in {}^1\mathcal{V}_\mathrm{h},$$

$$\left(\overline{A}_h \mathbf{u}_h, \mathbf{v}_h\right) = (\mathbf{u}_h, \mathbf{v}_h) + k\Theta a_h(\mathbf{u}_h, \mathbf{v}_h) + b_h(\mathbf{u}_h, \mathbf{u}_h, \mathbf{v}_h).$$

Then the Problem $(\mathcal{P})_\mathrm{m}$ can be rewritten as follows:

Problem $(\mathcal{P})_{\mathrm{m}'}$: Let $\mathbf{u}_h^0, \mathbf{u}_h^1, \ldots, \mathbf{u}_h^i$ be known. Then $\mathbf{u}_h^{i+\Theta}$ will be determined as

$$\begin{aligned}
&A_h\left(\mathbf{u}_h^{i+\Theta}, \mathbf{v}_h\right) + k\Theta b_h\left(\mathbf{u}_h^{i+\Theta}, \mathbf{u}_h^{i+\Theta}, \mathbf{v}_h\right) + \\
&+ k\Theta\left(\alpha_h \mathbf{u}_h^{i+\Theta}, \mathbf{v}_h\right) + k\Theta 2^{\frac{1}{2}} \hat{g}\left(m_h^{i+\Theta}, D\mathbf{v}_h\right) + \\
&+ k\Theta j_{gh}(\mathbf{v}_h) = \left(\mathbf{u}_h^i, \mathbf{v}_h\right) + k\Theta\left(\mathbf{F}_k^{i+\Theta}, \mathbf{v}_h\right) \quad \forall \mathbf{v}_h \in^1 V_h, \\
&\left(m - m_h^{i+\Theta}, D\mathbf{v}_h\right) \le 0 \\
&\forall m \in K = \left\{m | m_{jl} = m_{lj} \in L^\infty(\Omega \times I), \|m\|_\infty \le 1,\ j,l = 1 \ldots N\right\}, \\
&\qquad i = 0 \ldots n-1,\ \frac{1}{2} \le \Theta \le 1.
\end{aligned} \tag{95}$$

It is evident that the last condition is equivalent to that of the Problem $(\mathcal{P})_\mathrm{m}$ as for $m_h \in K_h$ we have

$$2^{\frac{1}{2}}(m_h, D\mathbf{v}_h) \le 2^{\frac{1}{2}} \int_\Omega (D_{ij}(\mathbf{v}_h) D_{ij}(\mathbf{v}_h))^{\frac{1}{2}}\, d\mathbf{x} = j_h(\mathbf{v}_h).$$

Next, let us introduce the Lagrangian $\mathcal{L}_h(\mathbf{v}_h, m_h)$ by

$$\mathcal{L}_h(\mathbf{v}_h, m_h) = \frac{1}{2}\left(\overline{A}_h\mathbf{v}_h, \mathbf{v}_h\right) + k\Theta(\alpha_h\mathbf{v}_h, \mathbf{v}_h) +$$
$$+k\Theta 2^{\frac{1}{2}}\hat{g}(m_h, D\mathbf{v}_h) + k\Theta j_{gh}(\mathbf{v}_h) - \left\langle L_h^i \mathbf{v}_h\right\rangle = \mathcal{J}(\mathbf{v}_h) + (m_h, \Psi_h(\mathbf{v}_h)),$$
$$\left\langle L_h^i, \mathbf{v}_h\right\rangle = \left(\mathbf{u}_h^i, \mathbf{v}_h\right) + k\Theta\left(\mathbf{F}_k^{i+\Theta}, \mathbf{v}_h\right), \quad \Psi_h(\mathbf{v}_h) = k\Theta 2^{\frac{1}{2}}\hat{g}D\mathbf{v}_h.$$

Then the Problem $(\mathcal{P})_{\mathrm{m}'}$ is equivalent to finding a saddle point of the Lagrangian $\mathcal{L}_h(.,.) : {}^1V_h \times K_h \to \mathbb{R}$ such that

$$\begin{aligned}
&\mathcal{L}_h\left(\mathbf{u}_h^{i+\Theta}, m_h^{i+\Theta}\right) \le \mathcal{L}_h\left(\mathbf{v}_h, m_h^{i+\Theta}\right) \qquad \forall \mathbf{v}_h \in {}^1V_h,\\
&\mathcal{L}_h\left(\mathbf{u}_h^{i+\Theta}, m_h\right) \le \mathcal{L}_h\left(\mathbf{u}_h^{i+\Theta}, m_h^{i+\Theta}\right) \qquad \forall m_h \in K_h,\\
&i = 0\ldots n-1, \ \frac{1}{2} \le \Theta \le 1.
\end{aligned} \tag{96}$$

To find a saddle point the Uzawa algorithm is used. To derive the discrete divergence $D_h\mathbf{u}_h = \mathrm{div}_h\mathbf{u}_h$ and the discrete approximation of pressure p the techniques of Temam (1979), that represent also an optimization problem based on the Uzawa algorithm, is used. Then both Uzawa's algorithms are compiled into one Uzawa's algorithm only (Hladik and Nedoma (1992), Nedoma (1998)) and the functional $j_{gh}(\mathbf{v}_h)$ will be approximated e.g., as $j_{gh}(\mathbf{v}_h) = j_{gh}\left(\mathbf{u}_h^i, \mathbf{v}_h\right)$ or by approximation of its regularized form.

The algorithm of thermal part of the problem, similarly as in the previous section, is parallel of that of Falleta and Lamichhane (2009).

Remark 10. *As was mentioned above the studied problem can be also solved by using the penalty and the regularization techniques. These both techniques can be also used for their numerical solutions.*

More information about special contact problems in different rheologies and the finite element methods can be found in Nečas and Hlaváček (1981), Panagiotopoulos (1985), Shi and Shillor (1992), Hlaváček et al. (1988), Hlaváček (2004), Hlaváček and Nedoma (2002), (2005), Nedoma (1983), (1987), (1993), (1997), (2000), (2001), (2003a), (2006b), (2013), Cocou and Scarella (2006) and about variational problems and about convex analysis see e.g., in Céa (1971), Ekeland and Temam (1976), Ionescu and Sofonea (1983), Ziedler (1990) and many others.

Acknowledgments

This work was supported by the longterm strategic development financing by the Institute of Computer Science ASCR v.v.i. (RVO:67985807). The author thanks to Dr. Ivan Hlaváček DrSc and Dr. Jiří Jarušek DrSc for numerous useful discussions about the contact problems. Moreover, the author thanks to Mrs Hana Bílková for her help with preparing of this chapter. My thanks go also to all from my institute and to all from the Nova Science Publishers who help me with preparation of this chapter presented in this book.

References

[1] K. Abe (1979). Size of the great earthquakes of 1837-1974 inferred from tsunami data. J. Geophys. Res. 84, 1561-1568.

[2] K. Abe (1981) . Physical size of tsunamigenic earthquakes of the northwestrn Pacific. Phys. Earth. Planet. Inter. 27, 194-205.

[3] R. A. Adams, *Sobolev Spaces*, Academic Press, New York, 1975.

[4] Aki, K., Richards, P.G. (1980). Quasistatic Seismology. Theory and Methods. Vol.1 and Vol.2, W.H. Freeman, San Francisco, 932 pp.

[5] K. Aki and P.G. Richards (2002). Quantitative Seismology. 2nd Edition, Univ. Science Books, Sansalito, California, 704pp.

[6] C. Atkinson (1979). Stress singularities and fracture mechanics. Appl. Mech. Rev. , 32, 123-135.

[7] B.K. Atkinson (1987). Introduction to fracture mechanics and its geophysical applications. In: B.K. Atkinson, Fracture Mechanics of Rocks. Academic Press, Geological Series, London, 1987.

[8] Bathe, K.J. (1996). Finite Element Procedures. Prentice-Hall, Englewood Cliffs, New Jersey.

[9] M. B*å*th (1979). Introduction to Seismology. Birkhäuser Vlg., Basel

[10] Belytschko, T., Liu W.K., Moran, B. (2002). Nonlinear Finite Elements for Continua and Structures. John Wiley&Sons, Ltd., Chichester.

[11] J. Berrocal and C. Fernandez (1996). Seismotectonic features of the Wadati-Benioff zone in the Andean region. In: Special Issue "Andean Tectonics", Third ISAG, St. Malo (France), 17-19/9/1996.

[12] J. Berrocal, J. Nedoma and Z. Kestřánek (2000). Numerical modelling of tectonic evolution of the Wadati-Benioff zone beneth the Andean region. Proc. of the 16th World IMACS Congress, Laussane.

[13] P. Bird (2003). An updated digital model of plate boundaries. Geochemistry, Geophysics, Geosystems, vol 4, Issue 3, March 2002, American Geophysical Union, 1027; doi:10.1029/2001GC000252,3.

[14] G. Blewitt et al. (2006). Rapid determination of earthquake magnitude using GPS for tsunami warning systems, Geophys. Res. Lett. , L11309, doi:10.1029/2006GL026145.

[15] P. Borman (Ed.) (2002). IASPEI New Manual of Seismological Observatory Practice. vol. 1, GeoForschungsZentrum Potsdam, Germany.

[16] S. Brunsson, F. Schmid, M. Schäfer, B. Wohlmuth (2007). A Fast and Robust Iterative Solver for Nonlinear Contact Problems Using a Primal-Dual Active Set Strategy and Algebraic Multigrid, Int. J. Numer. Math. Engng. 69, 524-543.

[17] E. Bryant (2001). Tsunami. The Underrated Hazard. Cambridge Univ. Press, New York.

[18] E. Bryant (2008). Tsunami. The Underrated Hazard. 2nd Edition Cambridge Univ. Press, New York.

[19] Capdeville, Y., Chaljub, E., Vilotte, J.-P., and Montagner, J.-P. (2003). Coupling the spectral element method with a modal solution for elastic wave propagation in global earth models. Geophys. J. Int., 152, 34-67.

[20] J. Céa, Optimisation, théorie et algorithms, Dunod, Paris, 1971.

[21] Chaljub, E., Capdeville, Y. and Vilotte, J.-P. (2003). Solving elastodynamics in a fluid solid heterogeneous sphere: a parallel spectral element approximation on non-conforming grids. J. Comput. Phys.,87, 457-491.

[22] M. Cocou, G. Scarella, G. (2006). Analysis of a Class of Dynamic Unilateral Contact Problems with Friction for Viscoelastic Bodies, In: Wriggers, P., Nackenhorst, V. (Eds), Analysis and Simulation of Contact Problems. Lecture Notes in Applied and Computational Mechanics, vol. 27, Springer Vlg, Berlin, Heidelberg, 137-144.

[23] K.C. Condie (1997). Plate Tectonics. 4th Edition. Butterworth-Heineman, Oxford, Amsterdam.

[24] B.N. Cox, H. Gao, D. Gross and D. Rittel (2005). Modern topics and challenges in dynamic fracture. J. Mech. Phys. Solids, 53, 555-596.

[25] F.A. Dahlen and J. Tromp (1998). Theoretical Global Seismology. Princeton University Press, Princeton, NJ.

[26] C. DeMets, R.G. Gordon and D.F. Argus (2010). Geologically current plate motions. Geophys. J. International, 181, 1-80.

[27] G. Duvaut, J. L. Lions (1976). Inequalities in Mechanics and Physics, Springer Vlg., Berlin, Heidelberg, New York.

[28] Ch. Eck, J. Jarušek, J. Krbec (2005). Unilateral Contact Problems. Variational Methods and Existence Theorems. Chapman & Hall/CRC, Taylor & Francis Group, Boca Raton, London, New York, Singapore, 2005.

[29] I. Ekeland, R. Temam (1976),(1979). Convex Analysis and Variational Problems. North-Holland, Amsterdam.

[30] S. Falletta and B.P. Lamichhane (2009). Mortar finite elements for a heat transfer problem on sliding meshes. Calcolo manuscipt, vol. 46, 131-148.

[31] E. E. Gdoutos. Fracture Mechanics. An Introduction. Springer Vlg., 2005.

[32] R. Glowinski (2003). Finite Element Methods in Incompressible Viscous Flow. In P.G. Ciarlet, J.L. Lions (Eds), Handbook of Numerical Analysis, Numerical Methods for Fluids (Part 3), vol. IX, Elsevier, Amsterdam.

[33] R. Glowinski, J. L. Lions, R. Trémolières, R.(1976). Analyse numérique des inéquations variationnelles. Part I and II, Dunod, Paris.

[34] R. Glowinski, J. L. Lions, R. Trémolières, R. (1981) Numerical Analysis of Variational Inequalities, North-Holland, Amsterdam.

[35] W. Han, S. Jensen and I. Shimanski (1997). The Kačanov method for some nonlinear problems. Appl. Numer. Math. 24,57-79.

[36] A. Hasegawa, N. Umio and A. Takagi (1978). Double-planed structure of the deep seismic zone in the nortwestern Japan arc. Tectonophysics, vol. 47, Issues 1-2, 43-58.

[37] J. Haslinger, I. Hlaváček, J. Nečas (1996). Numerical Methods for Unilateral Problems in Solid Mechanics, In: Handbook of Numerical Analysis, vol. IV, Elsevier, Amsterdam, 313-486.

[38] R. Hassani, I.R. Ionescu, T. Lachand-Robert (2005a). Optimization techniques in landslides modelling. Annals of Univ. of Craiova. Math. Comp. Sci Ser., vol. 32, 158-169.

[39] R. Hassani, I.R. Ionescu, T. Lachand-Robert (2005b). Shape optimization and supremal minimization approaches in landslides modelling. Applied Math. and Optimization, 52,3,349-364.

[40] G.P. Hayes and K.P. Furlong (2010). Quantifying potential tsunami hazard in the Puysegur subduction zone, south of New Zealand. Geophys. J.Int. 183, 1512-1524.

[41] P. Hild, I.R. Ionescu, T. Lachand-Robert and I. Rosca (2002). The bloking of an inhomogeneous Bingham fluid. Applications to landslides. Math. Model. Num. Anal. 36, 1013-1026.

[42] M. Hintermüller, K. Ito, K. Kunish (2003). The Primal-Dual Active Set Strategy as Semi-Smooth Newton Method. SIAM J. Optim., 13, 867-888.

[43] M. Hintermüller, V. Kovtunenko, K. Kunish (2004). The Primal-Dual Active Set Method for a Crack Problem with Non-Penetration, IMA J. Appl. Math. 69, 2004, 1-26.

[44] I. Hladík and J. Nedoma (1992). On the numerical solution of an initial boundary value problem for the visco-plastic Bingham fluid. In: M. Doležalová (Ed.). Numerical Methods in Geomechanics, Prague, 141-144.

[45] I. Hlaváček (2004). Mixed Finite Element Analysis of Semi-Coercive Unilateral Contact Problem with Given Friction. Tech. Report no 921, ICS AS CR, Prague.

[46] I. Hlaváček, J. Haslinger, J. Nečas, J. Lovíšek (1988). Solution of Variational Inequalities in Mechanics, Springer Vlg., New York.

[47] I. Hlaváček, J. Nedoma (2002). On a Solution of a Generalized Semi-Coercive Contact Problem in Thermo-Elasticity, Math. Comput. Simulat., 60, 1-17.

[48] I. Hlaváček, J. Nedoma (2005). Reliable Solution of an Unilateral Contact Problem with Friction and Uncertain Data in Thermo-Elasticity, Mathematics and Computers in Simulation 67, 559-580.

[49] S. Hüeber, B.I. Wohlmut (2005a). A Primal-Dual Active Set Strategy for Non-Linear Multibody Contact Problems, Comput. Meth. Appl. Mech. Engrg., 194, 3147-3166.

[50] S. Hüeber, B.I. Wohlmuth (2005b). An Optimal a Priori Estimate for Non-Linear Multibody Contact Problems, SIAM J. Numer. Anal., 43, 156-173.

[51] S. Hüeber, M. Mair, B.I. Wohlmuth (2005). A Priori Estimates and an Inexact Primal-Dual Active Set Strategy for Linear and Quadratic Finite Elements Applied to Multibody Contact Problems, Applied Numerical Mathematics, 54, 555-576.

[52] S. Hüeber, G. Stadler and B. Wohlmuth (2008). A primal-dual active set algorithm for three dimensional contact problems with Coulomb friction. SIAM J. Sci Comput., Vol. 30, No. 2, 572-596.

[53] I.R. Ionescu, M. Sofonea (1983). Functional and Numerical Methods in Viscoplasticity, Oxford Univ. Press, Oxford.

[54] B. Issacks and P. Molnar (1971). Distribution of stresses in the descending lithosphere from global survey of focal-mechanism solution of mantle earthquakes. Rev. Geophys. Space Phys. 9, 103-174.

[55] J. Jarušek (1983). Contact Problems with Bounded Friction. Coercive Case, Czechoslovak Math. J. 33(108), 1983, 237-261.

[56] J. Jarušek (1996). Dynamic Contact Problems with Given Friction for Viscoelastic Bodies, Czechoslovak Math. J. 46(121), 475-487.

[57] J. Jarušek, C. Eck (1999). Dynamic Contact Problems with Small Coulomb Friction for Viscoelastic Bodies. Existence of Solutions, Mathematical Models and Methods in Applied Sciences 9, 11-34.

[58] H. Kanamori and J.J. Cipar (1974). Focal process of the great Chilean earthquake May 22, 1960. Phys. Earth Planet Interiors, 9, 128-136.

[59] S. Karato, M.R. Reidel and D.A. Yuen (2001). Rheological structure and deformation of subducted slabs in the mantle transition zone: Implications for mantle circulation and deep earthquakes. Phys. Earth Planet. Inter., 127, 83-108.

[60] P. Kearey, K.A. Klepeis and F.J. Viese (2009). Global Tectonics. Wiley-Blackwell, Hoboken.

[61] H.B. Keating, C.F. Waythomas and A.G. Dawson (Eds) (2000). Landslides and Tsunamis. Birkhäuser Vlg, 1313 pp.

[62] D. Komatitsch and J. Tromp (1999). Introduction to the spectral-element method for 3D seismic wave propagation. Geophys. J. Int. 139, 806-822.

[63] D. Komatitsch and J. Tromp (2002a). Spectral-element simulations of global wave propagation. I. Validation. Geophys. J. Int., 149, 390-412.

[64] D. Komatitsch and J. Tromp (2002b). Spectral-element simulations of global wave propagation. II. 3-D models, oceans, rotation, and self-gravitation. Geophys. J. Int., 150, 303-318.

[65] D. Komatitsch and J. Tromp (2003). A Perfectly Matched Layer absorbing boundary condition for the second-order seismic wave equation. Geophys. J. Int., 154, 146-153.

[66] D. Komatitsch and J.-P. Vilotte (1998). The spectral-element method: an efficient tool to simulate the seismic response of 2D and 3D geological structures. Bull. Seismol. Soc. Am., 88(2), 368-392.

[67] D. Komatitsch, J.-P. Vilotte, R. Vai, J.K. Castillo-Covarrubias and F.J. Sánchez-Sesma (1999). The Spectral Element method for elastic wave equations: application to 2D and 3D seismic problems. Int. J. Numer. Meth. Engng., 45, 1139-1164.

[68] D. Komatitsch, C. Barnes and J. Tromp, J. (2000a). Simulation of anisotropic wave propagation based upon a spectral element method. Geophysics, 65(4), 1251-1260.

[69] D. Komatitsch, J. Ritsema and J. Tromp (2002b). The spectral-element method. Beowulf computing, and global seismology. Science, 298, 1739-1742.

[70] D. Komatitsch, C. Barnes, and J. Tromp (2000c). Wave propagation near a fluid-solid interface: a spectral element approach. Geophysics, 65, 623-631.

[71] D. Komatitsch, S. Tsuboi, C. Ji and J. Tromp (2003). A 146 bilion degrees of freedom, 5 teraflops, 2.5 terabyte earthquake simulation on the Earth Simulator. Proc. of ACM/IEEE Supercomputing SC'2003 conference, 2003, published on CD-ROM and at www.sc-conference.org/sc2003.

[72] Z. Kowalik and T.S. Murty (1993). Numerical Modelling of Ocean Dynamics. World Scientific Publ. , 481 pp.

[73] T.A. Laursen (2002), (2003). Computational Contact and Impact Mechanics. Springer, Berlin, Heidelberg.

[74] T. Lay and T.C. Walace (1995). Modern Global Seismology. Academic Press, San Diego, 521 pp.

[75] Liu, H.W. (1983). On the Fundamental Basis of Fracture Mechanics. Eng. Fracture Mech., 17, 5, 425-438.

[76] Liu, Q. (2006). Spectral-Element Simulations of 3-D Seismic Wave Propagation and Applications to Source and Structural Inversions. California Institute of Technology, Pasadena, California, PhD Thesis.

[77] C.L. Mader (2004). Numerical Modelling of Water Waves. CRC Press, 274 pp.

[78] T.W. McDevitt and T.A. Laursen (2000). A Mortar-Finite Element Formulation for Frictional Contact Problems, International Journal for Numerical Methods in Engineering, 48, 1525-1547.

[79] D.P. McKenzie and W.J. Morgan (1969). Evolution of triple junctions. Nature 224, 125-133.

[80] C. Miehe and P. Wriggers (1994). Contact constraints within coupled thermomechanical analysis - a finite element model. Computer Methods in Applied Mechanics and Engineering, 113, 301-319.

[81] D.C. Mosher et al. (Eds). Submarine Mass Movements and Their Consequences. Advanced in Natural and Technological Hazards Research, vol. 28, Springer Science +Business Media B.V.

[82] J. Nečas, J. Jarušek and J. Haslinger (1980). On the solution of the variational inequality to the Signorini problem with small friction. Boll. Unione Mat. Ital. 5, 17-B, 796-811.

[83] J. Nečas, I. Hlaváček (1981). Mathematical Theory of Elastic and Elasto-Plastic Bodies. An Introduction, Elsevier, Amsterdam, Oxford, New York, 1981.

[84] .J. Nedoma (1983). On one type of Signorini problem without friction in linear thermo-elasticity. Appl. Math. 28(6), 393-407.

[85] J. Nedoma (1987). On the Signorini problem with friction in linear thermo-elasticity: the quasi-coupled 2D-case, Appl. Math., 32, 3, 186-199.

[86] J. Nedoma, (1993). Finite Element Analysis in Nuclear Safety, Tech. Report V-550, ICS AS CR, Prague, and the short communication published in J.R. Whiteman (Ed). The Mathematics of Finite Elements and Applictions, John Wiley&Sons, Chichester, 1993, 385p.

[87] J. Nedoma (1995a). Equations of Magnetodynamics in Incompressible Thermo-Bingham's Fluid under the Gravity Effect, J.Comput. Appl. Math. 59, 109-128.

[88] J. Nedoma (1995b). Nonlinear Analysis of the Generalized Thermo-Magneto-Dynamic Problem, J.Comput. Appl. Math. 63, 1-3, 393-402.

[89] J. Nedoma (1997). On a Coupled Stefan-Like Problem in Thermo-Visco-Plastic Rheology, J.Comput. Appl. Math. 84, 45-80.

[90] J. Nedoma (1998). Numerical Modelling in Applied Geodynamics, John Wiley&Sons, New York, 966pp.

[91] J. Nedoma (1999). Analysis of a Coupled System of Equations of a Global Dynamic Model of the Earth, Mathematics and Computers in Simulation, 50, 265-283.

[92] J. Nedoma (2000). Finite Element Analysis of Elastic (Seismic) Wave Propagation in Region with Unilateral Contacts. In: Arabnia, H.R. (Ed.), Proc. Conference PDPTA'2000, Las Vegas, USA.

[93] J. Nedoma (2001). Dynamic Contact Problems with Friction in Elasticity and Thermo-Elasticity. In: Feistauer. M., Rannacher, R., Kozel, K. (Eds), Proc. of the 4th Conference on Numerical Modelling in Continuum Mechanics, MATFYZPRESS, Prague.

[94] J. Nedoma (2003a). Numerical Solution of a Stefan-like Problem in Bingham Rheology, Mathematics and Computers in Simulation, 61, 271-281.

[95] J. Nedoma (2003b). On a Coupled Thermo-Hydro-Mechanical Problem Based on the Thermo-Visco-Plastic Rheology. In: Proceedings of ICCSA-2003, Lecture Notes in Computer Science, vol. 2667, pp. 977-986, Springer, Berlin.

[96] J. Nedoma (2006a). On a Solvability of Contact Problems with Visco-Plastic Friction in the Thermo-Visco-Plastic Bingham Rheology, Future Generation Computer Systems, 22, 2006, 484-499.

[97] J. Nedoma (2006b). Security of Countries Endangered by Bigger Hurricanes: Proposal of the Research Project. Presented at the conference FEMTEC2006, El Paso, TX, USA and as an open letter to the Government of USA, 2005.

[98] J. Nedoma (2007). Tectonics along the Himalayan and Andean Mountain Ranges, Numerical Analyses of Collision and Subduction Zones, Technical Report No 1011, 2007.

[99] J. Nedoma (2010). Special problems in landslide modelling: Mathematical and Computational methods. In: E.D. Werner and H.P. Friedman (Eds). Landslides: Causes, Types and Effects. Nova Science Publishers, Inc., New York, 263-389.

[100] J. Nedoma (2012a). Mathematical modelling of some consequences of hurricanes: The proposal of research project, mathematical and computational methods. In: A. Tarasov and M. Demidov (Eds). Eddies and Hurricanes: Formation, Triggers and Impact. Nova Science Publishers, Inc., New York, 107-198.

[101] J. Nedoma (2012b). The Primal Dual Active Set (PDAS) method for dynamic variational inequalities arising from the fractured bone neoplasm models. TR No 1167, ICS AV CR, Prague.

[102] J. Nedoma (2013). Propagation of elastic waves in fractured media under a self-gravity field. TR-1183, ICS AS CR, Prague.

[103] J. Nedoma, L. Tomášek (2008). On a Solvability of Hydro-Mechanical Problem Based on Contact Problem with Visco-Plastic Friction in Bingham Rheology, J. Comput. Appl. Math., 218, 116-124.

[104] J. Nedoma, J. Stehlík, I. Hlaváček, J. Daněk, T. Doležalová, P. Přečková (2011). Mathematical and Computational Methods in Biomechanics of Human Skeletal System. An Introduction. Wiley, Hoboken, NJ, 572pp.

[105] J.G. Oldroyd (1947). A rational formulation of the equations of plastic flow for a Bingham solid. Proc. Camb. Philos. Soc. 43, 100-105.

[106] P. D. Panagiotopoulos (1985). Inequality Problems in Mechanics and Applications, Birkhäuser, Boston, Basel, Stuttgart.

[107] S.M. Peacock (2001). Are the lower planes of the double seismic zones caused by serpentinite dehydratation in subducting oceanic mantle? , Geology 29, 299-302.

[108] J.F. Price (2009). Metrics of hurricane-ocean interaction: Vertically-integrated or vertically-averaged ocean temperature. Ocean Sci Discuss 6, 909-951.

[109] J.R. Rice (1980). The Mechanics of Earthquake Rupture. In: Dziewonski, A.M., Boschi, E. (Eds). Physics of the Earth's Interior. Italian Physical Society/North-Holland, Bologna, Italy, 555-649.

[110] K. Satake (2012). Tsunamis generated by submarine landslides. In; Y. Yamada et al. (Eds). Submarine Mass Movements and Their Consequences. Springer Science, Berlin.

[111] K. Satake and Y. Tanioka (1999). Source of tsunami and tsunamigenic earthquakes in subduction zones. Pure and Appl. Geophys., 154, 467-483.

[112] P. Segal (1984). Rate-dependent extensional deformation resulting from crack growth in rock, J. Geophys. Res., 89(B6), 4185-4195.

[113] G. Seriani (1998). 3-D large-scale wave propagation modeling by spectral element method on Cray T3E multiprocessor, Comp. Meth. Appl. Mech. Eng., 164, 235-247.

[114] P. Shi, M. Shillor, (1992). Existence of a Solution to the n-Dimensional Problem of Thermo-Elastic Contact, Comm. PDE, 17, 1597-1618.

[115] P. Šolín, K. Segeth, and I. Doležal. (2004). Higher-Order Finite Element Methods, Chapman and Hall/CRC, Boca Raton.

[116] R. Temam (1977), (1979). Navier-Stokes Equations. Theory and Numerical Methods, North-Holland, Amsterdam.

[117] G.M. Thompson and J.R. Whiteman (1985). Analysis of strain representation in linear elasticity by both singular and non-singular finite elements. Numer. Methods Partial Differ. Eqs. 2, 85-104.

[118] S. Tsuboi, P.M. Whitmore and T.J. Sokolowski (1999). Application of M_{wp} to deep and teleseismic earthquakes. Bulletin of the Seismological Society of America, vol.89, 1345-1351.

[119] S. Uyeda (1982). Subduction zones: An introduction to comparative subductology. Tectonophysics, 81, 133-159.

[120] J.R. Whiteman and J.E. Akin (1984). Finite elements, singularities and fracture. In: Whiteman, J.R. (Ed.). The Mathematics of Finite Elements and Applications III. MAFELAP 1978, Academic Press, London, 35-51.

[121] B. Wohlmuth (2000). A mortar finite element method using dual spaces for the Lagrange multiplier. SIAM J. Numer. Qnal. 38, 989-1012.

[122] B. I. Wohlmuth, R. Krause (2003). Monotone Multigrid Methods on Nonmatching Grids for Non-Linear Multibody Contact Problems, SIAM J. Sci. Comput., 25, 1, 324-347.

[123] P. Wriggers (2002). Computational Contact Mechanics, John Wiley&Sons, Chichester.

[124] P. Wriggers, P.D. Panagiotopoulos (1999). New Developments in Contact Problems, Springer Vlg., Wien, New York.

[125] Y. Yamada et al. (2012). Submarine Mass Movements and Their Consequences. Springer Science, Berlin.

[126] E. Ziedler (1990). Applied Functional Analysis. Parts IIA and IIB, Vol 108-109, Appl. Math. Sci., Springer, New York.

In: Natural Disasters
Editor: Karina Linwood

ISBN: 978-1-63463-309-3

Chapter 3

WIND CHARACTERISTICS OF SEVERE TYPHOON AND WIND-INDUCED STRUCTURAL BEHAVIOR OF REINFORCED CONCRETE BUILDING IN BAYESIAN PERSPECTIVE

Sin-Chi Kuok and Ka-Veng Yuen*
Department of Civil and Environmental Engineering,
University of Macau, Macao, China

ABSTRACT

Severe typhoon generates violent aerodynamic excitation on coastal cities and they can induce destructive effects on the civil infrastructures. This chapter aims at investigating the wind characteristics of severe typhoon and the corresponding wind-induced effects on reinforced concrete building. For this purpose, full-scale monitoring on a 64.70m height reinforced concrete building was conducted during the passage of severe typhoon. An ultrasonic anemometer and geophone-based accelerometers were installed on the building to capture continuous measurements including the wind velocity and structural response. The measurements are utilized to identify the wind characteristic properties and evaluate the wind-induced structural effects. The analysis is carried out by employing the Bayesian probabilistic framework. Exploiting the full advantages of Bayesian inference, not only the optimal values of the concerned parameters can be estimated but the associated estimation uncertainty can also be quantified in the form of probabilistic distribution. Consequently, abnormal changes of the estimates can be distinguished explicitly from the variation due to statistical noise. The result provides a reliable basis for the understanding of the features of severe typhoons and diagnosing possible structural damage.

Keywords: Bayesian inference, typhoon, wind characteristics, wind-induced structural behavior, structural health monitoring

* Email: kvyuen@umac.mo. Tel: (853) 88224960. Fax: (853) 88222426 (Corresponding author).

1. Introduction

Full-scale monitoring under tropical cyclone condition provides valuable opportunity to investigate the aerodynamic behavior of severe wind flow and inspect the structural health condition of the instrumented structure. Considerable amount of research has been devoted to investigate the wind flow characteristic under severe tropical cyclone conditions and the structural behavior of landmarked infrastructures. Xu and Zhan (2001) conducted structural health monitoring on the 79-story Di Wang Tower and analyzed the wind measurements under severe typhoon [17]; Ni et al. (2007) addressed the quantification of structural performance of a cable-stayed bridge under both weak and strong wind condition [14]; Chen et al. (2011) presented the analytical and experimental modal analysis of the Guangzhou New TV tower under severe typhoons [7]; An et al. (2012) studied the wind characteristics of a severe typhoon on the supertall Shanghai World Financial Center [2]; Kuok and Yuen (2013) analyzed the wind features and structural performance of a residential building under severe typhoon Vicente, the strongest typhoon in the past decade experienced in the region [11]. All these previous studies showed that severe typhoon wind loading could induce temporary or even permanent effects to the attacked structures. Therefore, it is important to perform proper analysis to understand the behavior of severe tropical cyclone wind loading as well as the impacts on structural performance and structural health condition. In this chapter, the wind characteristic of severe typhoon and the corresponding wind-induced structural behavior are investigated in the Bayesian perspective.

Bayesian inference provides a rigorous framework for parametric identification [4, 16, 19, 27]. In contrast to deterministic parametric identification methodologies, Bayesian probabilistic approaches not only can determine the optimal estimates but also can quantify the associated estimation uncertainties in the form of probability distribution. By exploiting such promising feature, Bayesian inference has been successfully applied to various engineering disciplines, such as, air quality modeling [8], empirical relationships in soil mechanics [18], model class selection [5, 20, 25], outlier analysis [30], seismic attenuation [29], state-space analysis [28], and structural health monitoring [10, 22, 24]. Comprehensive review of Bayesian methods for structural dynamics and civil engineering can be found in [3, 19, 27]. In this study, the Bayesian probabilistic framework is employed to examine the in-field measurements of wind velocity and structural response captured under severe tropical cyclone loading. First, Bayesian model updating approach is developed to estimate the wind characteristic parameters. Thereafter, Bayesian spectral density approach [9, 21] is applied for the modal identification. Since the estimation uncertainty can be quantified, the fluctuation of the estimated parameters can be compared with the statistical uncertainty. Therefore, one can justify the significance of the wind-induced effects on the overall variation of the parameters.

This study aims to investigate the characteristic properties of severe tropical cyclone and the corresponding wind-induce structural performance. For this purpose, continuous monitoring on a 22-story reinforced concrete building was conducted under the passage of typhoon Utor. Typhoon Utor was the strongest tropical cyclone attacked Macau in 2013 and it was the only tropical cyclone necessitated the issuance of typhoon signal no. 8 by the Macao Meteorological and Geophysical Bureau in the same year. In-field measurements including wind velocity and structural response throughout the entire typhoon affected process were successfully captured. The Bayesian model updating approach and Bayesian spectral density approach are used to carry out the estimation of the wind characteristic parameters and modal parameters. The application demonstrates the efficacy and feasibility of the Bayesian probabilistic framework on parametric identification and uncertainty quantification. Moreover, the result provides useful analysis results to understand the aerodynamic behavior of severe wind flow and the induced wind structural interaction.

This chapter is organized as follows. Section 2 presents the formulation of the Bayesian approaches for the estimation of wind characteristic parameters and modal identification. Section 3 describes the instrumentation of the monitoring system. The information of the targeted building and the specifications of the monitoring instruments are introduced. Thereafter, Section 4 discusses the analytical results. The wind characteristic parameters, structural and excitation modal parameters as well as the wind-induced structural behavior are presented.

2. Bayesian Approaches for Parametric Estimation

Bayesian inference provides a promising and feasible probabilistic framework for parametric identification and uncertainty quantification. By applying the Bayesian framework, the Bayesian model updating approach for wind characteristic parameters estimation and the Bayesian spectral density approach for modal identification are developed. The formulation of these two Bayesian approaches is presented in the following sections.

2.1. Bayesian Model Updating for Estimation of Wind Characteristic Parameters

Bayesian model updating approach is introduced for the estimation of the wind characteristic parameters. Specifically, the wind spectrum density obtained from the longitudinal wind velocity is utilized as the data to estimate the turbulence integral length scale. The turbulence integral length scale is an important wind characteristic property to interpret the size of the large energy-containing eddies of the turbulence wind flow [6]. The widely adopted wind spectrum model, von-Karman wind spectrum model [6, 15], is employed to be the model class of the estimation. Detail formulation of the Bayesian model updating approach is presented in the following.

Use u and v to denote the longitudinal and lateral component of the measured wind velocity, respectively. Also, $\Delta\tau$ and T_τ are the sampling time interval and the duration of the measurement segment of the monitored wind velocity. For a measurement segment with $N_T = T_\tau/\Delta\tau$ samples, the von-Karman wind spectrum [6, 13, 15] is given by:

$$\frac{f_k S_u(f_k)}{\sigma_u^2} = \frac{4 f_k L_u/U}{[1+70.8(f_k L_u/U)^2]^{5/6}} \tag{1}$$

where S_u is the power spectral density of the longitudinal wind velocity; σ_u^2 is the variance of the longitudinal wind velocity; $U = \left(\sum \sqrt{u^2(t) + v^2(t)}\right)/N_T$ is the averaged wind speed in the horizontal plane; and L_u is the uncertain turbulence integral length scale; $\Delta f = 1/T_\tau$ is the frequency precision; and $f_k = k\Delta f, k = 0,1, \ldots, INT(N_\tau/2)$. Based on the von-Karman wind spectrum model, the measurement of the system at the kth frequency step can be expressed as:

$$y(k) = Q(k;\ L_u) + \epsilon(k) \tag{2}$$

where $y(k) = f_k S_u(f_k)/\sigma_u^2$ is obtained from the measured longitudinal wind velocity; and $Q(k) = 4 f_k L_u / \{U[1 + 70.8(f_k L_u/U)^2]^{(5/6)}\}$ expresses the relation between the fitting parameters and the measured quantity. The modeling error ϵ is modeled as zero-mean Gaussian independent and identical distribution (i.i.d.) process with variance σ_ϵ^2.

The uncertain parameter vector is taken to be $\boldsymbol{\theta} = [L_u, \sigma_\epsilon^2]^T$ that consists of the turbulence integral length scale and the variance of the modeling error. Assume that the uncertain parameters follow Gaussian distribution with mean $\widehat{\boldsymbol{\theta}}$ and covariance $\Sigma_{\boldsymbol{\theta}}$. Using the Bayes' theorem, the posterior probability density function (PDF) of the uncertain parameter vector is given by:

$$p(\boldsymbol{\theta}|D, C) = c_0 p(\boldsymbol{\theta}) p(D|\boldsymbol{\theta}, C) \tag{3}$$

where c_0 is the normalizing constant such that integrating the right hand side over the parameter space yields unity. The prior PDF $p(\boldsymbol{\theta})$ reflects the prior information of the parameters without the information of the data. The likelihood function $p(D|\boldsymbol{\theta}, C)$ reflects the contribution of the measured data D in establishing the posterior PDF of the parameters and it can be expressed as:

$$p(D|\boldsymbol{\theta}, C) = (2\pi)^{-\frac{N_f}{2}} |\boldsymbol{\Sigma}_\epsilon|^{-\frac{1}{2}} \exp\left\{-\frac{1}{2}\sum_{k=1}^{N_f} [y(k) - Q(k;\ L_u)]^T \boldsymbol{\Sigma}_\epsilon^{-1} [y(k) - Q(k;\ L_u)]\right\} \tag{4}$$

where $N_f = INT(N_\tau/2) + 1$ and $|.|$ is the determinant of a matrix.

Define the objective function as the negative logarithm of the posterior PDF, denoted as $J_1(\boldsymbol{\theta}) = -\ln[p(\boldsymbol{\theta}|D, C)]$. Therefore, the optimal model parameter vector $\widehat{\boldsymbol{\theta}}$ can be obtained by maximizing the posterior PDF which is equivalent to minimizing the objective function:

$$\widehat{\boldsymbol{\theta}} = \arg\min_{\boldsymbol{\theta}} J_1(\boldsymbol{\theta}) \tag{5}$$

Moreover, the covariance matrix of the optimal parameter vector can be computed by taking the inverse of the Hessian matrix evaluated at $\boldsymbol{\theta} = \widehat{\boldsymbol{\theta}}$:

$$\Sigma_{\boldsymbol{\theta}} = \left[\mathbf{H}(\widehat{\boldsymbol{\theta}})\right]^{-1} \equiv [\nabla J_1(\boldsymbol{\theta}) \nabla^T |_{\boldsymbol{\theta}=\widehat{\boldsymbol{\theta}}}]^{-1} \tag{6}$$

The covariance matrix quantifies the estimation uncertainty of the optimal model parameters. Therefore, rational judgment on the wind-induced variation of the estimates can be made. By applying the Bayesian model updating approach, the optimal value of the turbulence integral length scale and the modeling error as well as the associated estimation uncertainty can be obtained.

2.2. Bayesian Spectral Density Approach for Modal Identification

Bayesian spectral density approach is a well-developed probabilistic approach for modal identification [9, 21]. The modal identification is performed by utilizing the spectral density obtained from the structural response as the data to estimate the uncertain modal parameters. The efficacy and flexibility of the Bayesian spectral density approach on modal identification had been demonstrated through successful applications with in-field measurements [10, 26]. For completeness, the formulation of the approach is briefly reviewed.

Consider a linear dynamical system with N_d degrees of freedom:

$$\boldsymbol{M}\ddot{\boldsymbol{x}} + \boldsymbol{C}\dot{\boldsymbol{x}} + \boldsymbol{K}\boldsymbol{x} = \boldsymbol{T}_0\boldsymbol{F}(t) \tag{7}$$

where **M**, **C** and **K** are the mass, damping and stiffness matrix, respectively; and $\boldsymbol{T}_0$ is the force distributing matrix. The external excitation **F** can be modeled as zero-mean Gaussian white noise with spectral intensity matrix $\boldsymbol{S}_F(\omega) = \boldsymbol{S}_{F0}$. The measurement $\boldsymbol{Y}_N = \{\boldsymbol{y}(n), n = 1,2,\ldots,N\}$ contains N_o channels of structural response, corrupted by the measurement noise $\boldsymbol{\varepsilon}$:

$$\boldsymbol{y}(n) = \boldsymbol{L}_0\boldsymbol{q}(n) + \boldsymbol{\varepsilon}(n) \tag{8}$$

where $\boldsymbol{y}(n)\epsilon\mathbb{R}^{N_o}$ is the measurement at the nth time step; $\boldsymbol{q}(n)\epsilon\mathbb{R}^{N_d}$ is the concerned structural response (e.g., acceleration) at the same time step; and $\boldsymbol{L}_0\epsilon\mathbb{R}^{N_o\times N_d}$ is the observation matrix comprised of zeros and ones. The measurement noise $\boldsymbol{\varepsilon}$ is modeled as zero-mean Gaussian independent and identical distributed (i.i.d.) process with covariance matrix $\Sigma_{\boldsymbol{\varepsilon}}$.

Use $\boldsymbol{\alpha}$ to denote the uncertain modal parameter vector to be identified. It includes the structural modal parameters (i.e., modal frequencies, damping ratios and modal shape components) of the contributing modes and the characteristic parameters of the spectral intensity of the excitation and the measurement noise. To identify the uncertain modal parameter vector $\boldsymbol{\alpha}$, a discrete estimator of the power spectral density matrix $\boldsymbol{S}_{y,N}(\omega_k)$ is utilized [9]:

$$\boldsymbol{S}_{y,N}(\omega_k) = \frac{\Delta t}{2\pi N}\sum_{n,n'=0}^{N-1}\boldsymbol{y}(n)\boldsymbol{y}(n')^T\exp[-i\omega_k(n-n')\Delta t] \tag{9}$$

where Δt is the sampling time step; the superscript T denotes the transpose of a vector; $\Delta\omega = \frac{2\pi}{N\Delta t}$ is the frequency precision; and $\omega_k = k\Delta\omega, k = 0,1,\ldots,\mathrm{INT}(N/2)$.

With N_s independent sets of discrete time histories $\boldsymbol{Y} = \left\{\boldsymbol{Y}_N^{(s)}, s = 1,2, \dots, N_s\right\}$, the averaged spectral density matrix estimator can be obtained:

$$\boldsymbol{S}_{y,N}^{avg}(\omega_k) = \frac{1}{N_s}\sum_{s=1}^{N_s} \boldsymbol{S}_{y,N}^{(s)}(\omega_k) \tag{10}$$

where $\boldsymbol{S}_{y,N}^{(s)}(\omega_k)$ can be calculated using Eq. (9) with the measurement $\boldsymbol{Y}_N^{(s)}$. Given that $N_s \geq N_o$, the averaged spectral density matrix estimator follows the central complex N_o-variate Wishart distribution with N_s degrees of freedom. With a properly selected frequency range Ξ, the averaged spectral density matrix estimators in $\mathbf{S}_{\Xi}^{avg} = \left\{\boldsymbol{S}_{y,N}^{avg}(\omega_k), \omega_k \in \Xi\right\}$ are approximately independent [23].

Using the Bayes' theorem, the posterior PDF of $\boldsymbol{\alpha}$ given $\mathbf{S}_{\Xi}^{avg}$ is [4, 27]:

$$p(\boldsymbol{\alpha}|\mathbf{S}_{\Xi}^{avg}) = c_1 p(\boldsymbol{\alpha}) p(\mathbf{S}_{\Xi}^{avg}|\boldsymbol{\alpha}) \tag{11}$$

where c_1 is a normalizing constant. The prior PDF $p(\boldsymbol{\alpha})$ represents the prior information of the modal parameters in $\boldsymbol{\alpha}$.
Throughout this study, it is taken as a non-informative prior distribution so it can be absorbed into the normalizing constant. The likelihood function $p(\mathbf{S}_{\Xi}^{avg}|\boldsymbol{\alpha})$ is given by the product of Wishart distributions [9]:

$$p(\mathbf{S}_{\Xi}^{avg}|\boldsymbol{\alpha}) = \\ c_2 \prod_{\omega_k \in \Xi} \left|E\left[\boldsymbol{S}_{y,N}(\omega_k)|\boldsymbol{\alpha}\right]\right|^{-N_s} \exp\left[-N_s \mathrm{tr}\left(\left\{E\left[\boldsymbol{S}_{y,N}(\omega_k)|\boldsymbol{\alpha}\right]\right\}^{-1} \boldsymbol{S}_{y,N}^{avg}(\omega_k)\right)\right] \tag{12}$$

where c_2 is a constant that does not depend on the modal parameters; $E[.]$ is the mathematical expectation; tr(.) is the trace of a matrix, respectively. The optimal modal parameter vector $\widehat{\boldsymbol{\alpha}}$ can be determined by maximizing its posterior PDF.

To enhance the computational condition, the optimal modal parameters can be obtained equivalently by minimizing the objective function defined as $J_2(\boldsymbol{\alpha}) = -\ln p(\mathbf{S}_{\Xi}^{avg}|\boldsymbol{\alpha})$. Consequently, the covariance matrix of the modal parameters is given by the inverse of the Hessian of $J_2(\boldsymbol{\alpha})$ evaluated at $\boldsymbol{\alpha} = \widehat{\boldsymbol{\alpha}}$, i.e., $\Sigma_{\boldsymbol{\alpha}} = [\mathbf{H}(\widehat{\boldsymbol{\alpha}})]^{-1} \equiv [\nabla J_2(\boldsymbol{\alpha})\nabla^T|_{\boldsymbol{\alpha}=\widehat{\boldsymbol{\alpha}}}]^{-1}$ [19]. By employing the Bayesian spectral density approach with the measured structural acceleration response, the modal parameters and the associated estimation uncertainties can be determined.

3. Instrumentation of the Monitoring System

A monitoring system on the East Asia Hall was established to capture the full-scale measurements for the analysis. East Asia Hall was inaugurated in 2005 for athletes lodging in the 4th East Asian Games hosted in Macao. Figure 1 shows the side view and a typical floor plan of the building. It is a 22-story reinforced concrete building with L-shape floor layers.

The height of the building is 64.70 m, and the length of the two spans of the L-shape floor layers are 51.90 m and 61.75 m.

During the passage of typhoon Utor, continuous measurements on the wind velocity and structural response were successfully captured. The 3-dimensional wind velocity was captured with the Gill ultrasonic anemometer shown in Figure 2. The anemometer was installed on a 10 m height mast at the top of the building. The resolution of the measured wind velocity was 0.01 m/s and the sampling frequency was 32 Hz. On the other hand, a biaxial state-of-the-art accelerometer was placed at the junction of the two spans of the 18th floor to record the acceleration response of the building. The measured directions were indicated by the two perpendicular arrows shown in Figure 1(b). The accelerometer was operated under the standard exploration geophone spring-mass system with sensitivity of 50 V/g and sampling frequency of 200 Hz.

a

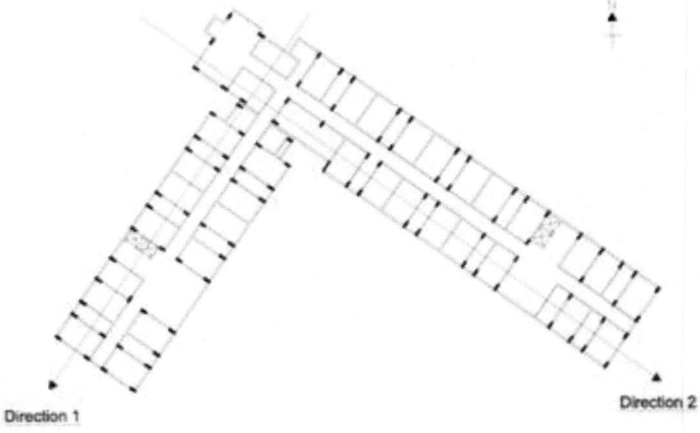

b

Figure 1. (a) Side-view and (b) typical floor plan of the East Asia Hall.

Figure 2. Photo of anemometer and its supporting mast.

4. Results

4.1. Meteorological Features

In this study, the monitoring wind velocity and structural response measured under the passage of typhoon Utor are examined. There were 31 tropical cyclones developed in the western Pacific Ocean during the 2013 Pacific typhoon season. However, typhoon Utor was the only tropical cyclone that necessitated the issuance of No. 8 typhoon signal by the Macao Meteorological and Geophysical Bureau in this year. It was an intense tropical cyclone that caused widespread damage across southeastern Asia. The tropical disturbance was formed on August 8 over the western North Pacific and it consolidated into Tropical Storm Utor in the next day. Thereafter, Utor intensified rapidly and developed its typhoon eye on August 10. As strengthened continuously over sea zone, Utor was developed into a severe typhoon in the morning of August 11 and became a super typhoon in the evening. Its highest 10-minute sustained near wind speed and lowest pressure reached 54.2 m/s and 925 hPa, respectively. On August 12, Utor crossed Luzon and weakened into a severe typhoon before entering the South China Sea. Moving gradually from the northern of the South China Sea, Utor passed the south side of Macau with the closest distance of 200 km in the morning of August 14.

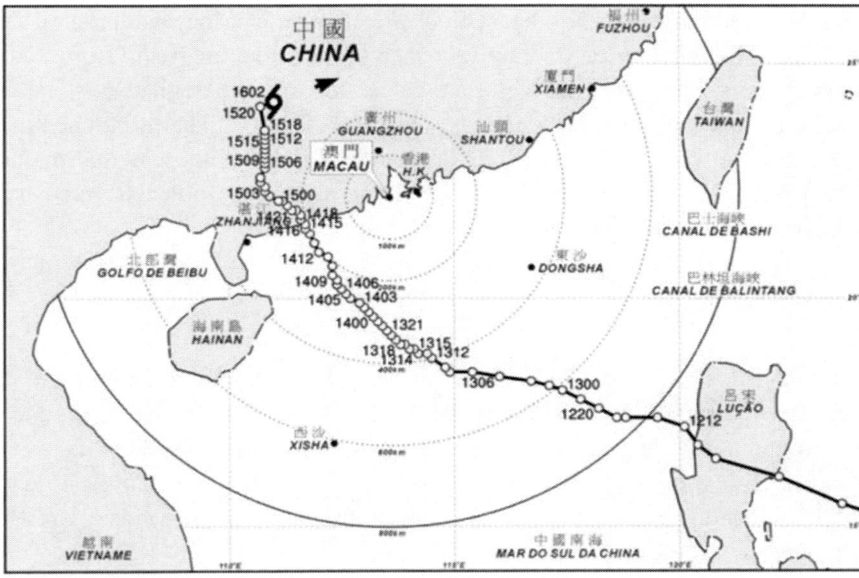

Figure 3. Track of typhoon Utor (provided by the Macao Meteorological and Geophysical Bureau (http://www.smg.gov.mo/www/e_index.php) and the time corresponding to GMT +08:00).

In the afternoon, it approached the coast of Guangdong and finally landed near Yangjiang. After made the landfall, the intensity of Utor decreased and finally dissipated over Guangxi on August 16. The track of typhoon Utor is shown in Figure 3. According to the severity of the wind conditions, typhoon signals were announced by the Macao Meteorological and Geophysical Bureau. The start and end time of the typhoon signal hoisted period were 18:00 of 12 August and 19:30 of 15 August, respectively. Throughout the entire 73.5 hours of monitoring period, continuous wind velocity and acceleration response were captured. The analysis results are presented in the following.

4.2. Analysis of Measured Wind Velocity

Figure 4 shows the time history of the 10-minute averaged horizontal wind speed and the associated wind direction. The dots represent the sustained wind speed and the crosses represent the wind direction. The peak of the sustained wind speed 18.9 m/s occurred at the 33th hour of the monitoring period. Before the peak occurs, the sustained wind speed increased gradually from the calm wind condition. After the wind speed reached its peak value, the sustained wind speed decreased and recovered to the level under calm wind condition. It is observed that the sustained wind speed of the wind weakening period (after the perk value) was more scattered than that of the wind strengthening period (before the peak value). This figure also shows the variation of the wind direction. Herein, the wind direction is expressed in the azimuth, i.e., counting clockwise from the true North. For the first ten hours, the wind direction changed suddenly from the direction of the prevailing wind (South) to the typhoon dominated direction (Northeast). The wind direction of the next twenty hours was relatively stable in the Northeast direction. During the 30th to 45th hours, the wind direction changes gradually from Northeast clockwise to South. Along with the dissipation of the typhoon effects, the wind direction returned back to the prevailing wind from South.

The turbulence intensity and gust factor are calculated to evaluate the statistical feature of the atmospheric turbulence and wind gustiness [1, 12-13, 15, 17]. The turbulence intensity is defined as the ratio between the standard deviation of the fluctuating wind and the horizontal mean wind speed. For each 10-minute time interval, the turbulence intensity of the longitudinal, lateral and vertical direction can be calculated by:

$$I_i = \frac{\sigma_i}{U}, i = u, v, w \tag{13}$$

where $i = u, v$ and w refers to the property of the longitudinal, lateral and vertical direction, respectively; σ_i is the standard deviation of the wind velocity of the i-direction and U is the 10-minute averaged horizontal mean wind speed. On the other hand, the gust factor quantifies the proportion of gust wind speed to the horizontal mean wind speed. The gust factor of the longitudinal, lateral and vertical direction for each 10-minute time interval is given by:

$$G_u = \frac{\max(\overline{u_{3s}})}{U}$$
$$G_v = \frac{\max(\overline{v_{3s}})}{U}$$
$$G_w = \frac{\max(\overline{w_{3s}})}{U} \tag{14}$$

Herein, the gust speed $\max(\overline{u_{3s}})$, $\max(\overline{v_{3s}})$ and $\max(\overline{w_{3s}})$ refer to the maximum 3-second sustained wind speed of the longitudinal, lateral and vertical wind component of the 10-minute time interval, respectively.

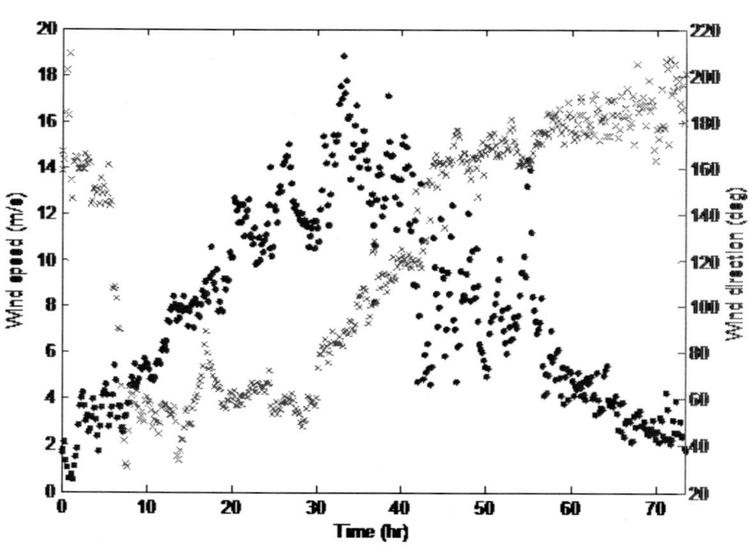

Figure 4. Time history of 10-minute averaged wind speed and wind direction.

The time histories of the calculated turbulence intensity and gust factor are shown in Figure 5 and 6, respectively. It is observed that the turbulence intensity and gust factor exhibited similar variation patterns. All the quantities were relatively stable during the wind strengthening period and the fluctuation became more severe during the wind weakening period.

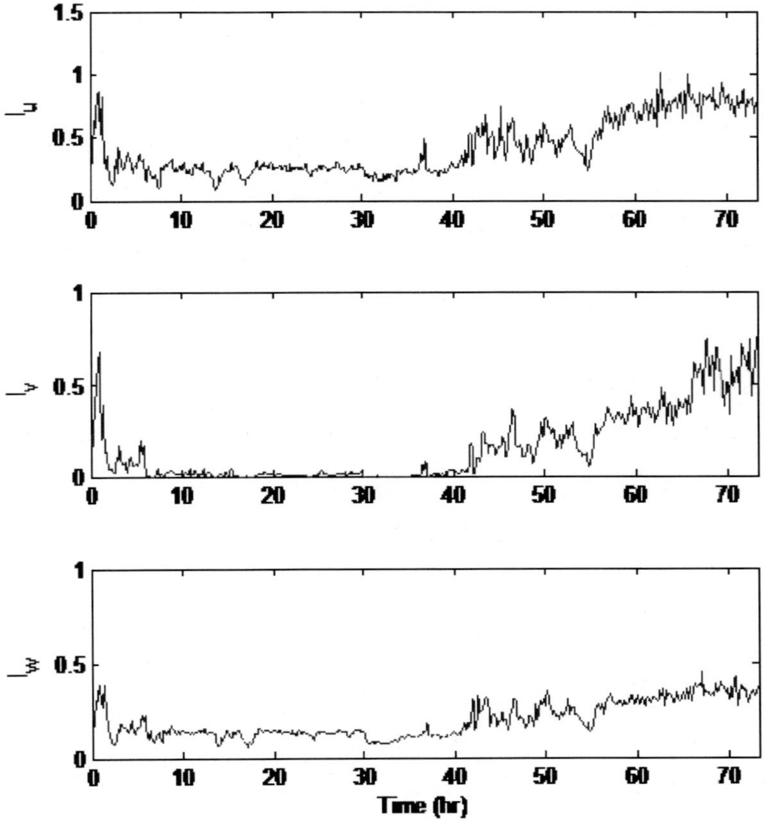

Figure 5. Time histories of turbulence intensity.

Moreover, the amplitudes of these quantities were generally higher in the wind strengthening period. By comparing the results of the longitudinal, lateral and vertical directions, it is found that the amplitude of both the turbulence intensity and gust factor of the longitudinal direction were the highest while the amplitude of the lateral direction were the lowest. Table 1 summarizes the minimum, maximum and average values of the turbulence intensity and gust factor of different directions. It shows that the averaged turbulence intensity of the longitudinal direction was more than two times of the values of other directions. Regarding the longitudinal direction, the maximum turbulence intensity was 1.01 which means that the standard deviation at this time instant was equal to the amplitude of the sustained wind speed.

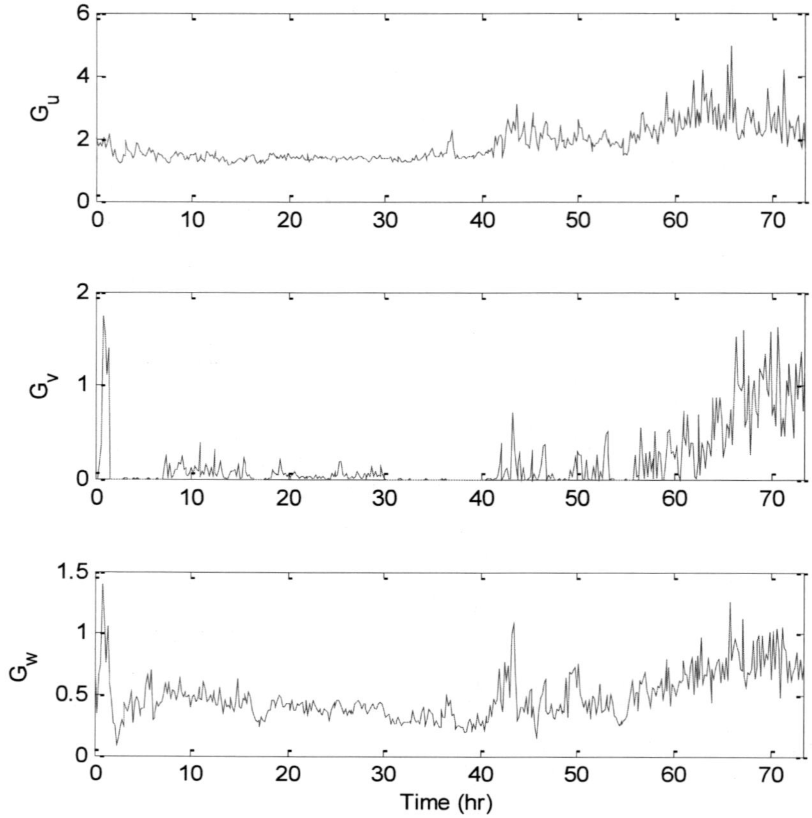

Figure 6. Time histories of gust factor.

Table 1. Minimum, maximum and average values of the turbulence intensity and gust factor

	Minimum	Maximum	Average
I_u	0.083	1.011	0.410
I_v	0	0.767	0.160
I_w	0.058	0.453	0.200
G_u	1.163	4.946	1.833
G_v	0	1.737	0.192
G_w	0.083	1.396	0.486

The maximum gust factor G_u was 4.95, indicating that the 3-second gust speed was almost five times higher than the corresponding sustained wind speed. The averaged gust factor of the longitudinal direction was more than three times of the value of other directions. For the minimum values, all of them occurred during the wind strengthening period. For the maximum values, the turbulence intensity of all the three directions and the gust factor of the longitudinal direction occurred during the wind weakening period.

Figure 7 shows the scatter plots of the turbulence intensity and gust factor versus the 10-minute averaged wind speed. The dots and crosses represent the data obtained from wind strengthening and weakening period, respectively. This notation is utilized for the rest of the scatter plots in this chapter. It is observed that the turbulence intensity and gust factor decreased as the wind speed increased for all the longitudinal, lateral and vertical directions. In other words, the wind velocity was less fluctuating with lower gustiness when the wind excitation was more severe. Moreover, the decreasing trends of the wind strengthening period and the wind weakening period were different.

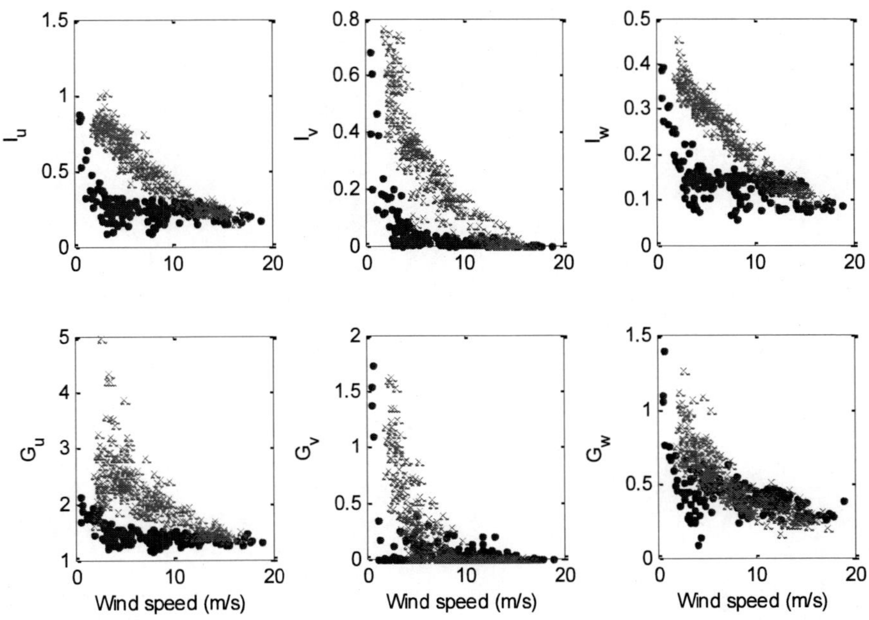

Figure 7. Wind characteristic properties versus 10-minute averaged wind speed.

For the data of the wind strengthening period, sudden drops of the quantities appeared in the low wind speed region. Thereafter, the data exhibited relatively flat decreasing trends. For the data of the wind weakening period, the data decreased gradually as the wind speed increased. Furthermore, the amplitude of the turbulence intensity and gust factor were relatively lower in the wind strengthening period even under the same wind speed. It indicates that during the wind strengthening period, the wind excitation was less fluctuated with lower gustiness.

The wind spectrum is constructed by utilizing the hourly data of the longitudinal wind velocity [13, 15]. Based on the wind spectrum obtained from the measurements, the turbulence integral length scale can be estimated using the Bayesian model updating approach. The wind spectrum corresponding to the peak sustained wind speed (the 33th hour) is shown in Figure 8. The fluctuating curve is obtained from the measurements and the smooth curve is the von-Karman wind spectrum with the turbulence integral length scale estimated using the Bayesian approach. It is found that this optimal model matches the trend of the wind spectrum obtained from the measurements. The estimation results of the turbulence integral length scale of the entire monitoring period are shown in Figure 9. In particular, Figure 9(a) shows the variation of the estimated values with the associated plus and minus three standard derivations confidence interval. The confidence interval encloses the region with a probability of 99.7%.

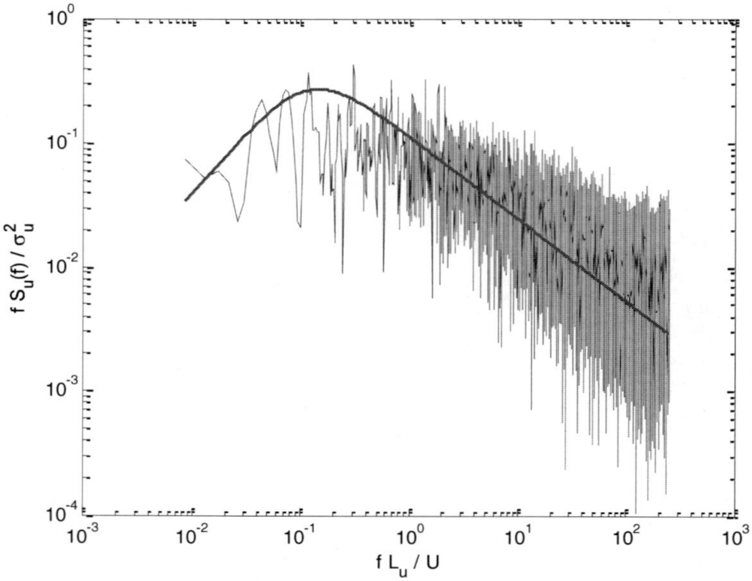

Figure 8. Wind spectrum with the fitting curve.

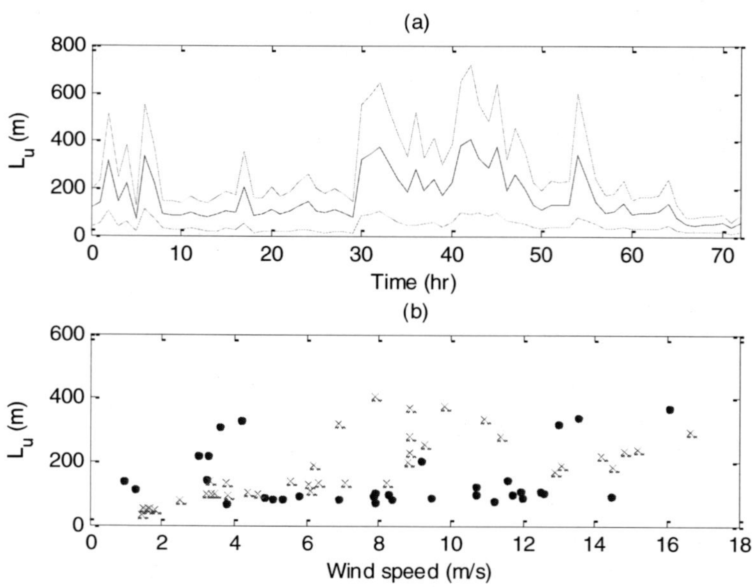

Figure 9. Turbulence integral length scale (a) estimation results and (b) scatter plot with hourly averaged wind speed.

The estimated turbulence integral length scale varied within 33.6 m and 403.5 m. Moreover, it is found that the confidence interval was relatively wide compares with the variation of the estimation results. For instance, the 99.7% probability confidence interval of the maximum turbulence integral length scale was [90.2, 716.8] m. On the other hand, Figure 9(b) shows the estimated turbulence integral length scale versus the hourly averaged wind speed. The data scattered over the entire region and no observable trend between the two quantities can be concluded.

4.3. Analysis of Measured Structural Response

The acceleration responses of the entire monitoring period are utilized for the modal identification. For each 10-minute segment of the acceleration responses, the Bayesian spectral density approach was implemented to estimate the modal parameters. In this study, the identification results of the first three modes are presented. Figure 10 shows the variation of the identified modal frequencies with the associated plus and minus three standard deviations confidence intervals. Since the building may not behave linearly under severe wind excitation, the identified structural modal parameters are referred to the corresponding quantities of the equivalent linear system and the reduction of the modal frequencies interprets the stiffness decrement of the equivalent linear system of the building.

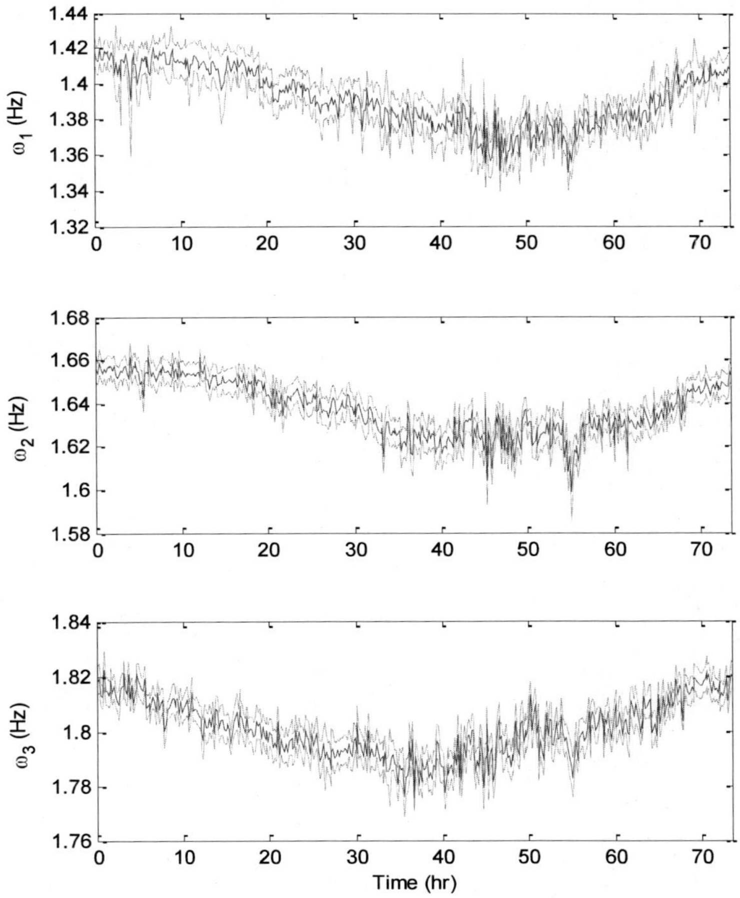

Figure 10. Identification results of the modal frequencies.

From the identification results, the modal frequencies of the three modes varied within the ranges of [1.340, 1.421] Hz, [1.599, 1.662] Hz and [1.776, 1.823] Hz. For these three concerned modes, the largest difference between the maximum and minimum value of the first mode was the largest that was equal to 5.24%. The correspond difference of the second and third mode are 3.94% and 2.66%, respectively. Moreover, the standard deviations of the estimates were less than 0.5% for all the three modes. Therefore, the confidence intervals were sufficiently narrow compared with the variations of the estimates.

On the other hand, the identified modal frequencies corresponding to calm wind condition were relatively higher than that corresponding to severe wind condition. Besides of the wind speed, other wind characteristic properties (such as wind direction and fluctuation) as well as structural properties (such as nonlinearity of construction materials, loading and unloading structural behavior) may also induce considerable effects on the structural performance [2, 7, 11, 13-14, 17]. However, such effects on different modes were not necessary the same. The results show that the variation of the modal frequencies of the third mode was mainly dominated by the wind speed. The corresponding modal frequencies showed a decreasing trend in the wind strengthening period and an increasing trend during the wind weakening period. However, the modal frequencies of the first two modes were more sensitive to other wind and structural factors. It turns out that the modal frequencies of these two modes presented decreasing trends until around the 45th hour which was twelve hours later than the occurrence of the peak wind speed.

Figure 11 shows the identified modal frequencies versus the 10-minute averaged wind speed. As expected, the data of the wind strengthening period were less scatter than those of the wind weakening period. The results were consistent with the observations shown in Section 4.2. Since the wind excitation was more fluctuated in the wind weakening period, the resultant structural modal parameters should be more scatter over the region. Nevertheless, distinct decreasing trends between the modal frequencies and sustained wind speed were observed in all the three modes. Moreover, the modal frequencies of the first two modes were generally higher in the wind strengthening period than those in the wind weakening period. However, the modal frequencies of the third mode obtained in the wind strengthening and weakening period were relatively closer. The data associated to the wind strengthening period were only slightly higher. The results show that the variation of the modal frequencies did not only depend on the strength of the sustained wind speed.

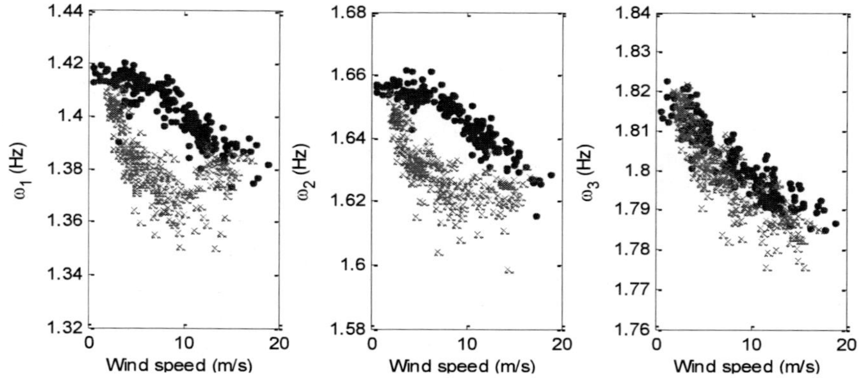

Figure 11. Identified modal frequencies versus averaged wind speed.

Finally, the energy of the wind-induced excitation is discussed. Figure 12 shows the time histories of the identified spectral intensities of the modal forces with the associated plus and minus three standard deviations confidence intervals. Herein, the quantities are normalized with the maximum value of the corresponding mode and thus the values vary from zero to unity. The normalized spectral intensities of the modal forces evaluate the excitation energy of a mode.

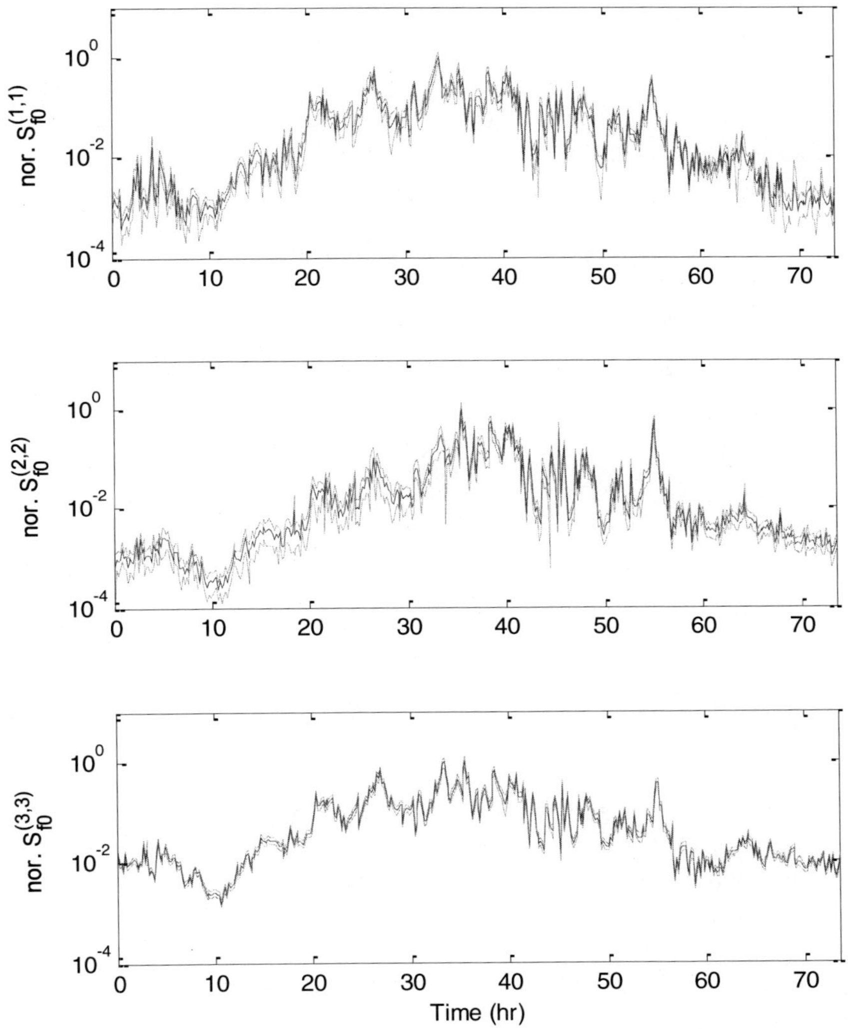

Figure 12. Identification results of the spectral intensities of the modal forces.

As shown in Figure 12, the spectral intensity of the modal force was significantly higher under severe wind loading. Specifically, the amplitude of the maximum values could be thousand times higher than the minimum values. The maximum spectral intensity of the first mode occurred at the 33rd hour of the monitoring time which was the same time of the peak of the sustained wind speed shown in Figure 4. On the other hand, the peaks of the second and third mode simultaneously at the 35th hour under sustained wind speed of 17.3 m/s. By comparing the results with Figure 10, it is clear that the spectral intensities presented opposite trends with the identified modal frequencies.

To illustrate the correlation between the energy of the excitation and the wind speed, Figure 13 shows the scatter plot between the identified spectral intensities of the modal forces and the 10-minute averaged wind speed. As expected, the identified spectral intensities increased as the sustained wind speed increases. The energy of the modal forces for the wind strengthening and weakening period were different even under the same sustained wind speed. In particular, the identified spectral intensities of wind strengthening period were generally lower than that of the wind weakening period. The results reconfirmed the conclusion drawn from Figure 11. Since the energy of the modal force in the wind weakening period was higher, the equivalent stiffness of the structural model and thus the resultant modal frequency was lower.

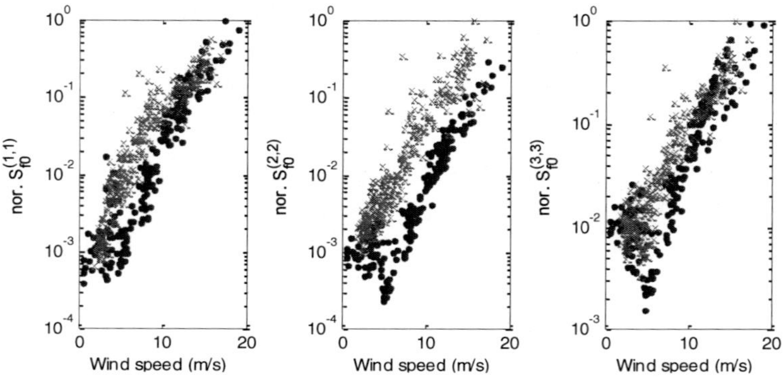

Figure 13. Identified spectral intensities of the modal forces versus the averaged wind speed.

CONCLUSION

In this study, full-scale monitoring measurements of a 22-story reinforced concrete building captured under the passage of typhoon Utor were examined. The monitored wind velocity and structural response were utilized to analysis the wind characteristic of tropical cyclone wind flow and the corresponding wind-induced structural behavior. First, basic wind characteristic parameters including sustained wind speed, wind direction, turbulence intensity and gust factor were calculated. Results showed that both turbulence intensity and gust factor exhibited decreasing trend with the sustained wind speed. Thereafter, the Bayesian model updating approach was utilized to estimate the turbulence integral length scale. For this parameter, no distinguishable trend with the wind speed could be observed. Finally, modal identification was performed by applying the Bayesian spectral density approach to the measured structural response. The identification results of the modal frequencies and the spectral intensities of the modal forces were presented. Result showed that the estimation uncertainties of these parameters were significantly small and the variation of these parameters was mainly caused by the wind-induced effects. It was found that the modal frequencies decreased as the sustained wind speed increased while the spectral intensities of the modal forces showed opposite tendencies. Moreover, it was observed that the contribution of the wind-induced effects on different modes were not necessary same. Results validated that the Bayesian approaches successfully estimated the wind characteristic parameters as well as the modal parameters. Furthermore, the quantified estimation uncertainty of the estimates provided important information for rational judgment on the significance of the wind-induced impact on the overall variation of the estimates.

ACKNOWLEDGMENT

This work was supported by the Science and Technology Development Fund of the Macau SAR government under Research Grant FDCT/012/2013/A1. This generous support is gratefully acknowledged.

REFERENCES

[1] Aly, A.M., Zasso, A. & Resta, F. (2011), Dynamics and Control of High-Rise Buildings under Multidirectional Wind Loads, *Smart Materials Research*.

[2] An, Y., Quan, Y. & Gu, M. (2012), Field Measurement of Wind Characteristics of Typhoon Muifa on the Shanghai World Financial Center, *International Journal of Distributed Sensor Networks,* 2012, 11.

[3] Beck, J.L. (2010), Bayesian System Identification Based on Probability Logic, *Structural Control & Health Monitoring,* 17(7), 825-847.

[4] Beck, J.L. & Katafygiotis, L.S. (1998), Updating Models and Their Uncertainties. I: Bayesian Statistical Framework, *Journal of Engineering Mechanics,* 124(4), 455-461.

[5] Beck, J.L. & Yuen, K.V. (2004), Model Selection Using Response Measurements: Bayesian Probabilistic Approach, *Journal of Engineering Mechanics,* 130(2), 192-203.

[6] Burton, T., Jenkins, N., Sharpe, D. & Bossanyi, E. (2011). *Wind Energy Handbook*: John Wiley & Sons.

[7] Chen, W., Lu, Z., Lin, W., Chen, S., Ni, Y., Xia, Y. & Liao, W. (2011), Theoretical and Experimental Modal Analysis of the Guangzhou New Tv Tower, *Engineering Structures,* 33(12), 3628-3646.

[8] Hoi, K.I., Yuen, K.V. & Mok, K.M. (2013), Improvement of the Multilayer Perceptron for Air Quality Modelling through an Adaptive Learning Scheme, *Computers and Geosciences,* 59, 148-155.

[9] Katafygiotis, L.S. & Yuen, K.V. (2001), Bayesian Spectral Density Approach for Modal Updating Using Ambient Data, *Earthquake Engineering & Structural Dynamics,* 30(8), 1103-1123.

[10] Kuok, S.C. & Yuen, K.V. (2012), Structural Health Monitoring of Canton Tower Using Bayesian Framework, *Smart Structures and Systems,* 10(4-5), 375-391.

[11] Kuok, S.C. & Yuen, K.V. (2013), Structural Health Monitoring of a Reinforced Concrete Building During the Severe Typhoon Vicente in 2012, *Scientific World Journal,* Article ID 509350.

[12] Li, Q., Xiao, Y., Wong, C. & Jeary, A. (2004), Field Measurements of Typhoon Effects on a Super Tall Building, *Engineering Structures,* 26(2), 233-244.

[13] Liu, H. (1990). *Wind Engineering: A Handbook for Structural Engineering*: Pearson Education.

[14] Ni, Y.Q., Ko, J.M., Hua, X.G. & Zhou, H.F. (2007), Variability of Measured Modal Frequencies of a Cable-Stayed Bridge under Different Wind Conditions, *Smart Structures and Systems,* 3(3), 341-356.

[15] Simiu, E. & Scanlan, R.H. (1996). *Wind Effects on Structures: Fundamentals and Applications to Design* (3rd ed.). New York: John Wiley.

[16] Vanik, M.W., Beck, J.L. & Au, S.K. (2000), Bayesian Probabilistic Approach to Structural Health Monitoring, *Journal of Engineering Mechanics,* 126(7), 738-745.

[17] Xu, Y. & Zhan, S. (2001), Field Measurements of Di Wang Tower During Typhoon York, *Journal of Wind Engineering and Industrial Aerodynamics,* 89(1), 73-93.

[18] Yan, W.M., Yuen, K.V. & Yoon, G.L. (2009), Bayesian Probabilistic Approach for the Correlations of Compressibility Index for Marine Clays, *Journal of Geotechnical and Geoenvironmental Engineering*, 135(12), 1932-1940.

[19] Yuen, K.V. (2010). *Bayesian Methods for Structural Dynamics and Civil Engineering*. Singapore; Hoboken, NJ: John Wiley & Sons.

[20] Yuen, K.V. (2010), Recent Developments of Bayesian Model Class Selection and Applications in Civil Engineering, *Structural Safety,* 32(5), 338-346.

[21] Yuen, K.V. & Beck, J.L. (2003), Updating Properties of Nonlinear Dynamical Systems with Uncertain Input, *Journal of Engineering Mechanics,* 129(1), 9-20.

[22] Yuen, K.V. & Katafygiotis, L.S. (2005), Model Updating Using Noisy Response Measurements without Knowledge of the Input Spectrum, *Earthquake Engineering & Structural Dynamics,* 34(2), 167-187.

[23] Yuen, K.V., Katafygiotis, L.S. & Beck, J.L. (2002), Spectral Density Estimation of Stochastic Vector Processes, *Probabilistic Engineering Mechanics,* 17(3), 265-272.

[24] Yuen, K.V. & Kuok, S.C. (2010), Ambient Interference in Long-Term Monitoring of Buildings, *Engineering Structures,* 32(8), 2379-2386.

[25] Yuen, K.V. & Kuok, S.C. (2010), Modeling of Environmental Influence in Structural Health Assessment for Reinforced Concrete Buildings, *Earthquake Engineering and Engineering Vibration,* 9(2), 295-306.

[26] Yuen, K.V. & Kuok, S.C. (2010), Monitoring a 22-Story Building under Severe Typhoons with Bayesian Spectral Density Approach, *Engineering Physics and Mechanics: Analysis, Prediction and Applications*, 509-533.

[27] Yuen, K.V. & Kuok, S.C. (2011), Bayesian Methods for Updating Dynamic Models, *Applied Mechanics Reviews,* 64(1) , Article number 010802.

[28] Yuen, K.V., Liang, P.F. & Kuok, S.C. (2013), Online Estimation of Noise Parameters for Kalman Filter, *Structural Engineering and Mechanics,* 47(3), 361-381.

[29] Yuen, K.V. & Mu, H.Q. (2011), Peak Ground Acceleration Estimation by Linear and Nonlinear Models with Reduced Order Monte Carlo Simulation, *Computer-Aided Civil and Infrastructure Engineering*, 26(1), 30-47.

[30] Yuen, K.V. & Mu, H.Q. (2012), A Novel Probabilistic Method for Robust Parametric Identification and Outlier Detection, *Probabilistic Engineering Mechanics,* 30, 48-59.

In: Natural Disasters
Editor: Karina Linwood

ISBN: 978-1-63463-309-3

Chapter 4

PHYTOPLANKTON RESPONSE IN RELATION TO THE PASSAGE OF TYPHOONS IN COASTAL REGIONS

Kenji Tsuchiya** and Tatsuki Toda
Soka University, Tangi-machi, Hachioji-shi, Tokyo, Japan

ABSTRACT

Strong wind and heavy rain accompanied by typhoon passage induce physical disturbances such as upwelling, vertical mixing, terrestrial runoff and sediment resuspension, which supply nutrients to the photic layer in coastal waters. Consequently, phytoplankton biomass and production increase significantly. Primary production increased up to 19 times compared to that of pre-typhoon condition. Assimilation number (ratio of primary production to chlorophyll *a* concentration), which is an indicator of phytoplankton activity, is enhanced significantly after the passage of typhoon. In a temperate bay, a high assimilation number lasted for nine days after the passage of typhoon. The integrated primary production during the nine days after the passage of typhoon accounted for 7.2-9.1% of the annual primary production in the upper waters of the bay. Approximately three typhoons approach the bay annually, which implies that primary production in the upper waters of the bay may be enhanced in the range of 21.6-27.3% of the annual primary production.

One of the major fates of a phytoplankton bloom is the sinking of cells from the euphotic zone to deep waters. Typhoon-induced phytoplankton bloom is characterized by dominance of diatoms, such as *Skeletonema* spp., *Chaetoceros* spp., *Leptocylindrus* spp. and *Pseudo-nitzschia* spp. As diatoms show high sinking rates, their blooms may enhance the sinking flux of biogenic carbon to deep waters after the passage of typhoon. Another factor to consider is micro- and meso-zooplankton grazing on phytoplankton.

An abrupt decline in phytoplankton biomass and increase in zooplankton fecal pellets after the passage of typhoon suggests termination of the bloom due to intense zooplankton grazing.

In this chapter, the impact of typhoon on the phytoplankton biomass, production and species composition, and the fate of phytoplankton in coastal waters is documented and the ecological response of phytoplankton to the passage of typhoon is discussed.

---

* Corresponding author: Kenji Tsuchiya. Soka University, 1-236, Tangi-machi, Hachioji-shi, Tokyo 192-8577, Japan. E-mail: ktsuchiya@soka.gr.jp.

INTRODUCTION

Phytoplankton is a major organism at the base of aquatic food web. The abundance and growth, as well as biogeochemistry, can be influenced by regular seasonal variations of biotic and abiotic factors, and irregularly and strongly influenced by climatic and anthropogenic perturbations in coastal waters (Paerl et al. 2006). Coastal regions account for only 7.6% of the ocean's surface area, but they contribute approximately 30% of the global primary production (Sverdrup et al. 1942, Liu et al. 2000). However, changes of biogeochemistry and food web induced by these perturbations are complex, and our knowledge of how phytoplankton and coastal environments are influenced by different types of perturbations remains fragmented.

Typhoons, including hurricanes and tropical cyclones, generally occur in the ocean with > 27 ºC of sea surface temperature (SST), and consistently high temperatures contribute to keeping typhoon intensities. Such high SST is observed in tropical and subtropical regions throughout the year, and in temperate regions during summer, where the water column become strongly stratified. In stratified ocean environments, nutrients are depleted in the upper waters, and phytoplankton biomass and production are limited by nutrients although the supply of light energy is abundant. The passage of typhoons induces physical disturbances such as terrestrial runoff (Zheng and Tang 2007, Chen et al. 2009), sediment resuspension (Fogel et al. 1999), and vertical mixing and upwelling (Price 1981, Lin et al. 2003, Chung et al. 2012) in coastal waters. These disturbances load a large amount of nutrients into the euphotic zone (Table 1). For example, NO_2+NO_3, PO_4 and $Si(OH)_4$ concentrations increased ~30-fold, ~38-fold and ~6.8-fold, respectively, after the passage of typhoon. These nutrient loadings lead to large phytoplankton blooms and shifts in phytoplankton species composition (e.g., Fogel et al. 1999, Shiah et al. 2000, Chen et al. 2009). Understanding the fate of phytoplankton blooms induced by typhoon passage may contribute to clarifying the ecological role of typhoon. For example, when increased phytoplankton biomass is grazed by zooplankton, the fixed carbon and energy is consequently transferred to organisms in higher trophic levels such as fishes. In other words, typhoons may enhance fish production and play an important role in fisheries. On the other hand, phytoplankton blooms are mainly formed by diatoms (e.g., Chung et al. 2012), which tend to sink into the deeper waters as their strategy to retrieve nutrients (Smayda 1997). Therefore, most of the increased phytoplankton cells would sink before being grazed by zooplankton, enhancing the function of the biological pump that transports and stores carbon fixed at upper waters into deep waters.

In this chapter, the impact of typhoon on the phytoplankton biomass, production and species composition, and the fates of phytoplankton in coastal waters are documented and the ecological response of phytoplankton to the passage of typhoon is discussed.

BIOMASS AND PRODUCTION OF PHYTOPLANKTON

Accompanied by typhoon passage, physical disturbances such as terrestrial runoff, sediment resuspension, vertical mixing and upwelling load nutrients into the euphotic zone, significantly increasing phytoplankton productivity (e.g., Shiah et al. 2000). For example, in Sagami Bay, concentrations of NO_2+NO_3, PO_4 and $Si(OH)_4$ increased after the passage of typhoons, and they were significantly higher than the background concentrations ($P < 0.05$, Mann-Whitney U-test, Figure 1; Tsuchiya et al. 2014). Chlorophyll *a* (chl *a*) concentrations increased and reached their maximums 4 to 6 days after the passage of typhoons (Figure 1). Chl *a* concentration and primary production before and after typhoon passages are summarized in Table 2. The ratios (A/B) of various values in the literature after typhoon passages (A) and either before typhoon passages or the average values during non-typhoon periods (B) provides a good assessment on the impact of typhoons on phytoplankton biomass and production.

Table 1. Summary of nutrients before and after typhoon passages (B: Before typhoon, A: After typhoon)

Nutrients	Study site	Typhoon/Tropical cyclone	Concentration (μM)			References
			B	A	A/B	
NO_2+NO_3	South Carolina	*Dennis*	0.10	2.1	21	Zeeman 1985
	Northern Taiwan	*Tim*	-	12	-	Chang et al. 1996
	Northern Taiwan	*Caitlin, Doug*	2.0	51	26	Chang et al. 1996
	Northern Taiwan	*Fred*	7.0	6.0	0.9	Chang et al. 1996
	North Carolina	*Gordon*	~0.3	1.0	3.3~	Fogel et al. 1999
	Taiwan Strait	*Herb*	~0.5	0.60-5.3	1.2-11	Shiah et al. 2000
	Philippine Sea	*Pabuk, Wutip, Sepat*	0.013	0.39	30	Chen et al. 2009
	Northeast of Taiwan	*Morakot*	~0.5	5.1	10~	Chung et al. 2012
	Sagami Bay	*Mawar*	4.2*	18	4.3	Tsuchiya et al. 2014
	Sagami Bay	*Sinlaku*	4.2*	17	4.0	Tsuchiya et al. 2014
	Sagami Bay	*Etau*	4.2*	10	2.4	Tsuchiya et al. 2014
	Sagami Bay	*Malou*	4.2*	18	4.3	Tsuchiya et al. 2014
PO_4	South Carolina	*Dennis*	0.06	0.21	3.5	Zeeman 1985
	Northern Taiwan	*Tim*	-	0.53	-	Chang et al. 1996
	Northern Taiwan	*Caitlin, Doug*	0.28	0.52	1.9	Chang et al. 1996
	Northern Taiwan	*Fred*	0.20	0.32	1.6	Chang et al. 1996
	North Carolina	*Gordon*	~0.1	0.27	2.7~	Fogel et al. 1999
	Philippine Sea	*Pabuk, Wutip, Sepat*	0.013	0.045	3.5	Chen et al. 2009
	Northeast of Taiwan	*Morakot*	~0.1	0.30	3~	Chung et al. 2012
	Sagami Bay	*Mawar*	0.04*	1.5	38	Tsuchiya et al. 2014
	Sagami Bay	*Sinlaku*	0.04*	0.54	14	Tsuchiya et al. 2014
	Sagami Bay	*Etau*	0.04*	0.70	18	Tsuchiya et al. 2014
	Sagami Bay	*Malou*	0.04*	1.0	25	Tsuchiya et al. 2014
$Si(OH)_4$	Northern Taiwan	*Tim*	-	20	-	Chang et al. 1996
	Northern Taiwan	*Caitlin, Doug*	9.0	19	2.1	Chang et al. 1996
$Si(OH)_4$	Northern Taiwan	*Fred*	2.5	17	6.8	Chang et al. 1996
	Northeast of Taiwan	*Morakot*	2.1-2.6	12	4.6-5.7	Chung et al. 2012
	Sagami Bay	*Mawar*	18*	69	3.8	Tsuchiya et al. 2014
	Sagami Bay	*Sinlaku*	18*	45	2.5	Tsuchiya et al. 2014
	Sagami Bay	*Etau*	18*	59	3.3	Tsuchiya et al. 2014
	Sagami Bay	*Malou*	18*	62	3.4	Tsuchiya et al. 2014

*The background data (non-typhoon periods) were obtained from the monthly sampling of surface water for the months from July to September (during typhoon seasons) from 2008 to 2010 in Sagami Bay.

The A/B ratios of chl *a* concentration and primary production reported by previous studies were 0.3–32 and 0.4–19, respectively. While most of the previous studies showed an increase in chl *a* concentration and primary production (e.g., Chen et al. 2009), a few studies reported a decrease in chl *a* concentration and primary production after the passage of typhoon (e.g., Zeeman 1985, Shiah et al. 2000, McKinnon et al. 2003).

Zeeman (1985) suggested that the decrease in chl *a* concentration after the passage of typhoon was the result of terrestrial water containing lower chl *a* concentrations. Figure 1(d, l and p) shows an example of temporary decrease in chl *a* concentration just after the passage of typhoon. There are a limited number of studies conducting *in situ* daily samplings before and after the passage of typhoon since it is dangerous and difficult to predict the track of typhoon (Chung et al. 2012, Tsuchiya et al. 2013, Tsuchiya et al. 2014). Daily samplings were not conducted in Zeeman (1985), Shiah et al. (2000) and McKinnon et al. (2003) and they would likely miss a phytoplankton bloom if it was very transient and shorter than their sampling interval (Delesalle et al. 1993). The ratio of primary production to chl *a* concentration (PP/Chl *a* ratio or assimilation number) is an indicator of phytoplankton photosynthetic activity. In Sagami Bay, PP/Chl *a* ratio showed relatively high values until 9 days after the passage of typhoon *Mawar* (131 ± 29 mg C [mg Chl $a]^{-1}$ d^{-1}; Tsuchiya et al. 2013, Figure 2) and until 7 days after the passage of typhoon *Malou* (117 ± 37 mg C [mg Chl $a]^{-1}$ d^{-1}; Tsuchiya 2014, Figure 2).

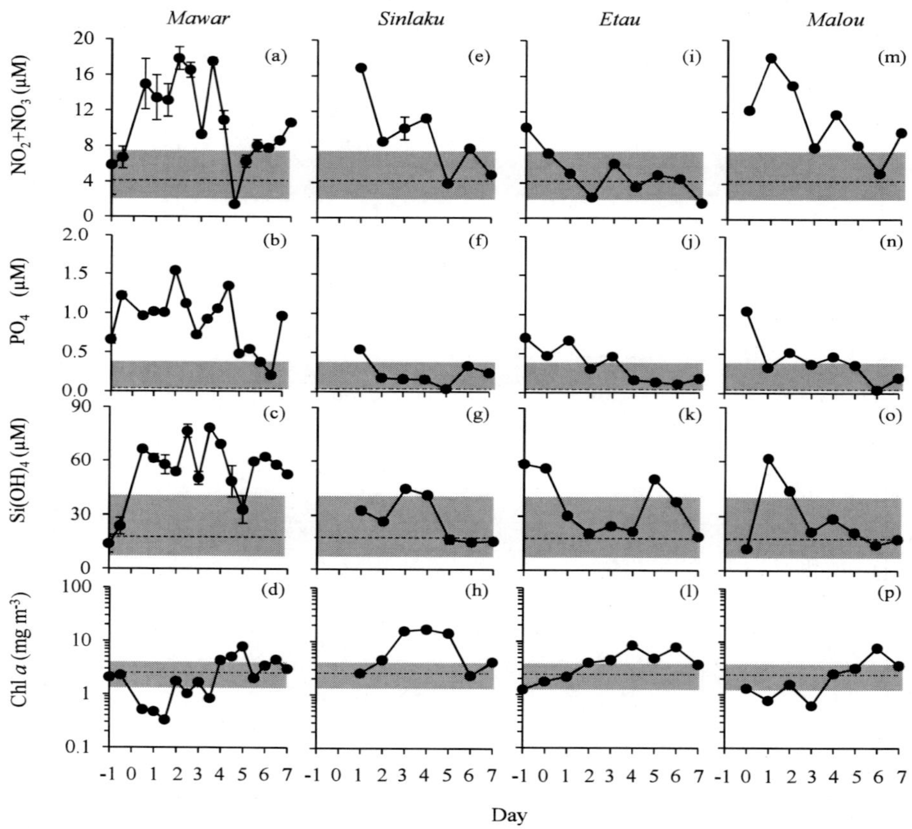

Modified from Tsuchiya et al. 2014.

Figure 1. Temporal variations in NO_2+NO_3, PO_4, $Si(OH)_4$ and chl *a* concentrations in (a-d) *Mawar*, (e-h) *Sinlaku*, (i-l) *Etau* and (m-p) *Malou* at Manazuru Port in Sagami Bay, Japan. Dashed lines and gray shadows indicate 50% percentiles (median) of the background data (non-typhoon) and percentile contours at the 5% and 95% levels, respectively.

Table 2. Summary of chlorophyll *a* concentration and primary production before and after typhoon/tropical cyclone passages

Study site	Typhoon/ Tropical cyclone	Chlorophyll *a* concentration			Primary production			References
		Before (B)	After (A)	A/B	Before (B)	After (A)	A/B	
		$mg\ m^{-2}$			$g\ C\ m^{-2}\ d^{-1}$			
Southern Northwest Shelf of Australia	*Tiffany*	6.6-79	16-25	0.3-3.2	1.3-8.9	2.8-3.4	0.4-2.2	McKinnon et al. 2003
		$mg\ m^{-3}$						
South China Sea	*Kai-Tak*	0.10	3.2	32	0.3	2.8	9.3	Lin et al. 2003
South China Sea	*Ling-Ling*	0.14	0.45	3.2	0.27	2.6	9.6	Zhao et al. 2008
South China Sea	*Kai-Tak*	0.15	0.56	3.7	0.31	4.1	13	Zhao et al. 2008
South China Sea	*Hagibis*	0.14	0.74	5.3				Sun et al. 2010
					$mg\ C\ m^{-3}\ d^{-1}$			
South Carolina	*Dennis*	1.0-11	2.5-6.0	0.56-2.6	0.13-2.1×10^3	0.83-2.5×10^3	1.2-6.3	Zeeman 1985
The Reef of Tiafura	*Wasa*	0.12	0.21	1.8	9.1	14	1.5	Delesalle et al. 1993
Northern Taiwan	*Tim*	0.40-1.8	4.2	2.3-11				Chang et al. 1996
Northern Taiwan	*Caitlin, Doug*	0.30	3.5	12				Chang et al. 1996
Northern Taiwan	*Fred*	0.60	2.0	3.3				Chang et al. 1996
North Carolina	*Gordon*	0.56	0.56-3.4	1.0-6.0	18	53	2.9	Fogel et al. 1999
Taiwan Strait[a]	*Herb*	0.28-0.48	0.42-0.91	0.89-2.1	2.2-2.6	10-50	3.9-19	Shiah et al. 2000
Pamlico Sound	*Dennis, Floyd, Irene*	4.8	5.6-20.7	1.2-4.3				Paerl 2001
Gulf of Mexico	*Ivan*	0.24-0.36	0.81-0.99	2.3-4.1				Walker et al. 2005
South China Sea	*Damrey*	0.15	0.72	4.8				Zheng and Tang 2007
Neuse River Estuary	*Helene*	20.6	24.5	1.2				Wetz and Paerl 2008
Neuse River Estuary	*Isabel*	27.1	31.1	1.1				Wetz and Paerl 2008
		$mg\ m^{-3}$			$mg\ C\ m^{-3}\ d^{-1}$			
Neuse River Estuary	*Alex, Bonnie, Charley*	8	17.3	2.2				Wetz and Paerl 2008
Philippine Sea[a]	*Pabuk, Wutip, Sepat*	0.06-0.11	0.22-0.56	2.0-5.6	7.6-8.0	12-76	1.5-10	Chen et al. 2009
Northeast of Taiwan	*Morakot*	0.3-0.4	3.7	9.3-12				Chung et al. 2012
Sagami Bay	*Mawar*	2.6	7.8	3.0	80[b]	349	4.4	Tsuchiya et al. 2013
Sagami Bay	*Sinlaku*	1.9[c]	17	8.9				Tsuchiya et al. 2014
Sagami Bay	*Etau*	1.2	8.4	6.8				Tsuchiya et al. 2014
Sagami Bay[a]	*Malou*	1.9[c]	7.7-8.0	4.0-4.2	80[b]	512-561	6.4-7.0	Tsuchiya 2014

[a]Ranges of data are based on individual calculations from multiple sampling stations; [b]Primary production before typhoons was the mean value in summer derived from Satoh et al. (2000); [c]Chl*a* concentration before typhoons was the mean value in summer derived from Sugawara et al. (2003) and Miyaguchi et al. (2006).

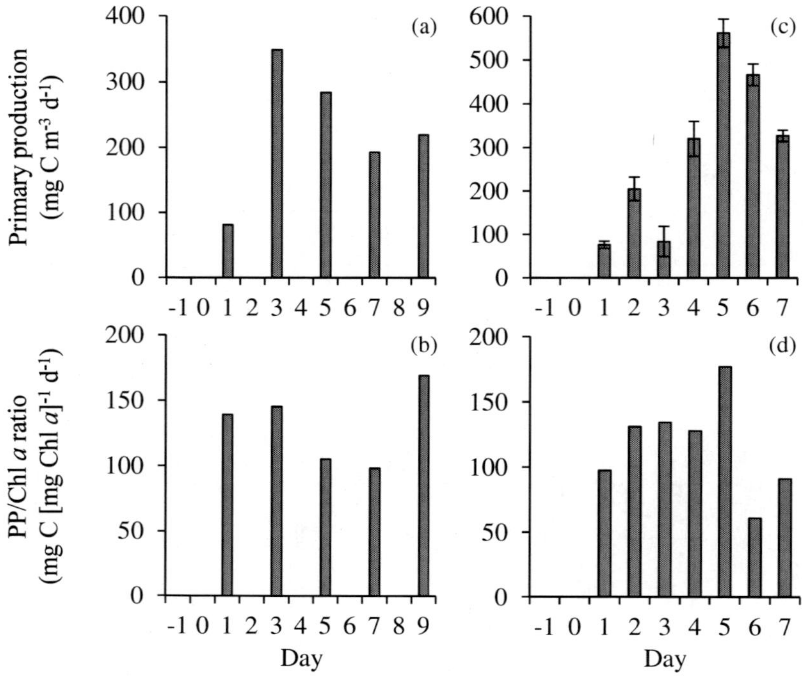

Modified from Tsuchiya et al. 2013 and Tsuchiya 2014.

Figure 2. Temporal variations in primary production and the ratio of primary production to chl *a* concentration (PP/Chl a ratio) in (a-b) *Mawar* and (c-d) *Malou* at Manazuru Port in Sagami Bay, Japan.

These PP/Chl *a* ratios were significantly higher than the mean PP/Chl *a* ratio of 40 mg C [mg Chl $a]^{-1}$ d^{-1} reported by Sugawara et al. (2003) in August and September in Sagami Bay (Student's t-test, $p < 0.01$).

There were significant positive correlations between PP/Chl *a* ratio, and NO_2+NO_3, PO_4 and $Si(OH)_4$ concentration after the passage of typhoons (Tsuchiya 2014, Figure 3).

The results suggest that phytoplankton responded rapidly to nutrient injection, and primary productivity was significantly enhanced by the passage of typhoons.

The integrated primary productions for the duration of nine days after the passage of *Mawar* and for the duration of seven days after the passage of *Malou* were 2.10×10^3 mg C m $^{-3}$ and 2.07×10^3 mg C m^{-3}, which accounts for 7.1–9.1% of the annual primary production in the upper waters of Sagami Bay [annual primary production values were derived from Satoh et al. 2000 (2.92×10^4 mg C m^{-3} yr^{-1}), Sugawara et al. 2003 (2.30×10^4 mg C m^{-3} yr $^{-1}$) and Ara and Hiromi 2009 (2.82×10^4 mg C m^{-3} yr^{-1})].

Approximately 3 typhoons on average approach Sagami Bay annually (Japan Meteorological Agency), which implies that primary production in the upper waters of Sagami Bay may be enhanced in the range of 21.3–27.3% of the annual primary production.

These results verify that episodic events such as typhoons can make large contributions to the annual primary production in coastal regions.

In the oceanic region of the South China Sea, integrated primary production estimates at the bloom center after typhoon passages accounted for 9.9–23% of the annual primary production (Liu et al. 2002, Lin et al. 2003, Zhao et al. 2008).

Table 3. Major phytoplankton species in the pre- and post-typhoon passage (B: Before typhoon, A: After typhoon)

Location	Typhoon	Phytoplankton species		References
		B	A	
Puerto Rico	*Edith*	-	*Skeletonema* spp.	Glynn et al. 1964
Chesapeake Bay	*Agnes*	Nanoplankton	*Gymnodinium nelsoni*	Loftus and Seliger 1976
South Carolina	*Dennis*	*Skeletonema* spp.	*Skeletonema* spp.	Zeeman 1985
		Rhizosolenia alata	*Rhizosolenia alata*	
		Thallasiothrix sp.	*Pleurosigma* sp.	
		Ondontella autita	*Coscinodiscus* sp.	
			Ceratium tirchoceros	
			Gymnodinium sp.	
			Dinophysis caudata	
Great Barrier Reef	-	-	*Chaetoceros curvisetum*	Furnas 1989
			Nitzschia sp.	
			Small pennate diatoms	
Reef of Tiafura	*Wasa*	-	*Nitzschia closterium*	Delessalle et al. 1993
			Nitzschia bicapitata	
			Pennate diatoms	
Northern Taiwan	*Tim, Caitlin, Doug, Fred*	-	*Skeletonema* spp.	Chang et al. 1996
			Chaetoceros sp.	
			Nitzschia sp.	
Kaneohe Bay	-	-	*Protoperidinium* spp.	Hoover et al. 2006
			Ceratium spp.	
			Guardia spp.	
			Chaetoceros spp.	
Neuse River Estuary	*Isabel*	Dinoflagellates	Diatoms	Valdes-Weaver et al. 2006
Neuse River Estuary	*Helene*	Dinoflagellates	Dinoflagellates	Wetz and Paerl 2008
			Cyptophytes	
Neuse River Estuary	*Alex, Bonnie, Charley*	Cyanobacteria	Cyanobacteria	Wetz and Paerl 2008
		Tricodesmium sp.	Dinoflagellates	
Philippine Sea	*Pabuk, Wutip, Sepat*	*Tricodesmium* sp.	*Hemiaulus* spp.	Chen et al. 2009
		(Unicellular cyanobacteria)	*Guinardia* sp.	
			Chaetoceros spp.	
Northeaast of Taiwan	*Morakot*	*Trichodesmium* spp.	*Chaetoceros* spp.	Chung et al. 2012
		Gymnodinium spp.		
Sagami Bay	*Mawar*	*Skeletonema* spp.	*Skeletonema* spp.	Tsuchiya et al. 2014
		Dinophysis spp.	*Protoperidinium* spp.	
			Chaetoceros spp.	
Sagami Bay	*Sinlaku*	-	Dinoflagellates	Tsuchiya et al. 2014
			Chaetoceros spp.	
Sagami Bay	*Etau*	*Chaetoceros* spp.	*Ceratium* spp.	Tsuchiya et al. 2014
			Protoperidinium spp.	
			Chaetoceros spp.	
Sagami Bay	*Malou*	-	*Ceratium* spp.	Tsuchiya et al. 2014
			Protoperidinium spp.	
			Prorocentrum spp.	
			Dinophysis spp.	
			Chaetocero spp.	
			Erataulina spp.	

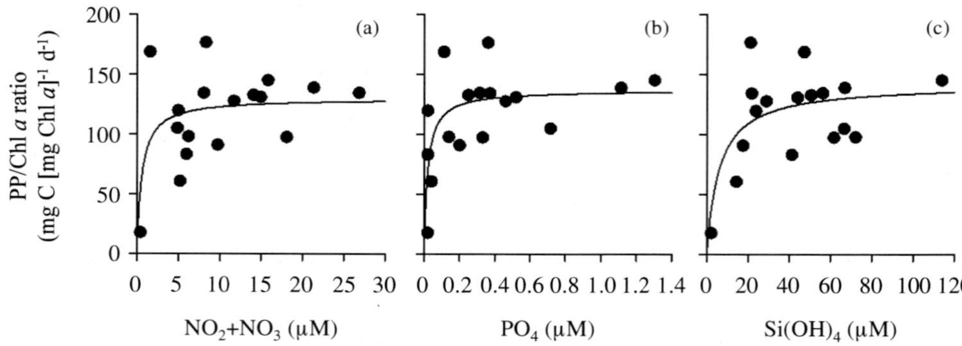

Modified from Tsuchiya 2014.

Figure 3. Relationships between PP/Chl *a* ratio, and (a) NO_2+NO_3 concentration and ([PP/Chl *a* ratio] = 130 × [NO_2+NO_3] / ([NO_2+NO_3] + 0.58); $r = 0.53$, $n = 17$, $P < 0.05$), (b) PO_4 concentration ([PP/Chl *a* ratio] = 137 × [PO_4] / ([PO_4] + 0.02); $r = 0.62$, $n = 17$, $P < 0.05$) and (c) $Si(OH)_4$ concentration ([PP/Chl *a* ratio] = 142 × [$Si(OH)_4$] / ([$Si(OH)_4$] + 6.0); $r = 0.66$, $n = 17$, $P < 0.01$) after the passage of *Mawar* and *Malou* at Manazuru Port and the fixed offshore station in Sagami Bay, Japan.

PHYTOPLANKTON SPECIES COMPOSITION

After the passage of typhoons, diatoms and dinoflagellates generally become the dominant taxa in phytoplankton communities (Table 3). Relatively high proportions of dinoflagellates occurred just after the passage of typhoons in Chesapeake Bay (Loftus and Seliger 1976), South Carolina (Zeeman 1985), Kaneohe Bay of Hawaii (Hoover et al. 2006) and Sagami Bay (Tsuchiya et al. 2014).

For example, dinoflagellates such as *Protoperidinium* spp. and *Ceratium* spp. dominated the phytoplankton community in a week after the storm at low salinity levels (Hoover et al. 2006).

At Manazuru Port in Sagami Bay, dinoflagellates were dominant at low salinity just after the passage of typhoons (Figure 4). Moreover, at a fixed offshore station (depth 120 m) in Sagami Bay, a noticeable dinoflagellate surface maximum was observed following an abrupt decline in surface salinity just after the passage of typhoon *Malou*, (Figure 5; Tsuchiya 2014).

Typhoon-driven runoff may contribute to the formation of a shallow, low salinity, and nutrient-rich fresh water layer at the surface, as suggested by Satoh et al. (2000). It is known that the swimming behavior of dinoflagellates could serve to concentrate cells into the nutrient-rich fresh water layer at the surface (Donaghay and Osborn 1997). Besides, harmful algal blooms are frequently formed in the pycnocline of coastal systems after heavy rainfall, calm sunny weather, influxes of estuarine waters, or interactions between dissimilar water masses (Ryther 1955). Hoover et al. (2006) suggested that the early appearance of dinoflagellates after episodic events is consistent with observations (Sellner et al. 2001) that they were better adapted to utilize high-nutrient concentrations under low-turbulence conditions than other phytoplankton groups. Dinoflagellates are also known to utilize organic substrates for nutrition (Zeeman 1985), which would favor production in waters enriched with organic matter through typhoon-driven terrestrial runoffs.

In fact, dominant dinoflagellates after such episodic events were often *Protoperidinium* spp., *Ceratium trichoceros*, *Ceratium* spp., *Prorocentrum* spp., *Dinophysis caudata*,

Dinophysis spp. and *Gymnodinium* spp. (Zeeman 1985, Hoover et al. 2006, Tsuchiya et al. 2014), and these species are considered heterotrophic or mixotrophic dinoflagellates (Lessard and Swift 1985, Hansen 1991, Strom and Morello 1998, Jacobson 1999, Jeong 1999, Morton 2000, Ismael 2003, Nishitani et al. 2003, Sherr and Sherr 2007).

Thus, these physical (i.e., sharp pycnocline at the surface induced by runoff) and biological features (i.e., their swimming or physiological characteristics) may have allowed the relative success of dinoflagellates just after the passage of typhoons.

Diatoms dominated the phytoplankton bloom after the passage of typhoon in almost all regions (Table 3). For instance, a significant positive relationship between total phytoplankton biomass and relative contribution of diatom was observed in Sagami Bay after the passage of typhoons ($P < 0.05$, Figure 6). In Kaneohe Bay of Hawaii, after the dominance of *Protoperidinium* spp., the composition of phytoplankton shifted from a mixed assemblage of the pennate diatoms *Haslea* spp., *Pleurosigma* spp. and *Thalassionema nitzschioides* and the dinoflagellate *Ceratium* spp., to the centric chain diatoms *Guardia* spp. and *Chaetoceros* spp. (Hoover et al. 2006). In Puerto Rico, *Skeletonema costatum* dominated up to 99% of the phytoplankton abundance after typhoon *Edith* (Glynn et al. 1964). In the southern East China Sea, the diatom population was dominated by chain-forming centric diatoms (*Chaetoceros* spp.) instead of the *Trichodesmium* spp. and *Gymnodinium* spp. that prevailed before the typhoon (Chung et al. 2012). Also, the dominant phytoplankton shift from cyanobacteria such as *Trichodesmium* sp. to diatoms such as *Hemiaulus* spp., *Guinardia* sp. and *Chaetoceros* spp. was observed in the Philippine Sea after the passage of typhoons *Pabuk*, *Wutip* and *Sepat* (Chen et al. 2009). Diatoms are known to require less essential trace metals (Ho, 2006), have efficient carbon fixation mechanisms like the C-4 pathway which functions as a CO_2 concentrating mechanism (Reinfelder et al., 2000, 2004) and have large central vacuoles in which nutrients can be stored for later use (Tozzi et al., 2004). During *in situ* experiments, Furnas (Furnas, 1989, 1991) observed that the difference between maximal growth rates of diatom species and flagellate populations was of the order of 1-3 doublings day^{-1}. As a result of those characteristics, diatoms have an advantage over dinoflagellates under conditions where nutrients are available from episodic loading even though dinoflagellates were temporary dominant just after the passage of typhoons (Furnas, 1990; Raven, 1997; Katz et al., 2004; Tozzi et al., 2004).

In the diatom community, *Skeletonema* spp. and *Chaetoceros* spp. were the most dominant taxa after the passage of typhoons in Sagami Bay (Figure 7). However, they did not appear concurrently; *Skeletonema* spp. were dominant after the passage of *Mawar*, while *Chaetoceros* spp. were dominant after the passages of *Sinlaku*, *Etau* and *Malou* (Tsuchiya et al. 2014).

Phytoplankton succession is known to be linked to and regulated by the variability in nutrients (Escaravage et al. 1996, Rees et al. 1999, Estrada et al. 2003, Lagus et al. 2004). In order to assess the competition mechanism between the two phytoplankton taxa within the community, the relationship between their dominance and nutrient stoichiometry were examined. A significant relationship was found between the relative dominance of each taxa, and their respective N/P ratios ($P < 0.05$; Figure 7), whereby the contribution of *Skeletonema* spp. biomass was higher under low N/P ratios while that of *Chaetoceros* spp. was lower under the high N/P ratio. The N/P ratios after the passage of typhoons were 9.9 in South Carolina (Zeeman 1985) and 11.5 in northern Taiwan (Chang et al. 1996), which are similar to the mean value (12.4) found after the passage of *Mawar* when *Skeletonema* spp. dominated the community. Roden and O'Mahony (1984) conducted an outdoor experiment in an enclosure and reported that *Skeletonema* spp. favor low N/P ratios and were dominant when the N/P ratio was less than 30. Meanwhile, the growth of *Chaetoceros* spp. was not limited by phosphorus when the N/P ratio was more than 30. Alkaline phosphatase activity of *Chaetoceros affinis* was higher than that of *Skeletonema costatum*, especially under high N/P ratio conditions (Møller et al. 1975). These previous studies support the idea that *Chaetoceros* spp. could dominate the phytoplankton community under higher N/P ratios, or lower PO_4 concentrations. These relationships support the classic interpretation where the N/P ratio determines the dominant phytoplankton bloom (e.g., Tilman et al. 1986, Escaravage et al. 1996, Rees et al. 1999, Estrada et al. 2003, Lagus et al. 2004). In Sagami Bay, the contribution of nutrient sources might be different between *Mawar* and the other three typhoons; the dominant nutrient inputs were likely from terrestrial runoff and upwelling in *Mawar*, and only from terrestrial runoff in *Sinlaku*, *Etau* and *Malou*. Terrestrial waters supply a large amount of nitrogen to coastal regions (Edmond et al. 1985, Turner et al. 1990, Fujiki et al. 2004), which leads to higher N/P ratios compared to that of deep water in Sagami Bay (Kamatani et al. 2000). The results suggest that the proportional contribution of nutrient sources, which determines nutrient stoichiometry after the passage of typhoons, is an important factor in controlling phytoplankton community succession.

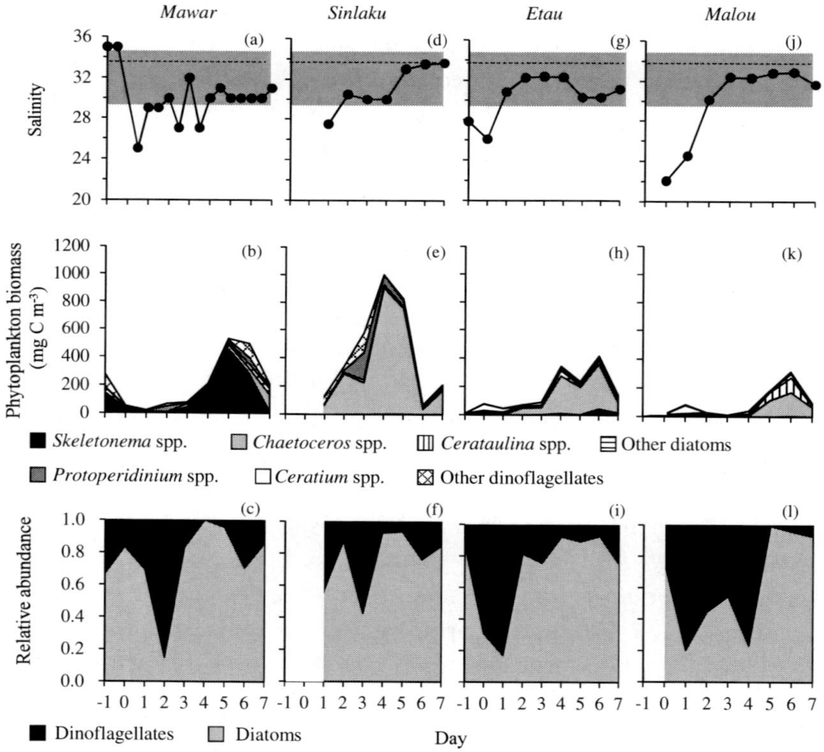

Modified from Tsuchiya et al. 2014.

Figure 4. Temporal variations in salinity, phytoplankton biomass and relative abundance of diatom and dinoflagellate in (a-c) *Mawar*, (d-f) *Sinlaku*, (g-i) *Etau* and (j-l) *Malou* at Manazuru Port in Sagami Bay, Japan. Dashed lines and gray shadows indicate 50% percentiles (median) of the background data (non-typhoon) and percentile contours at the 5% and 95% levels, respectively.

THE FATE OF PHYTOPLANKTON AFTER THE PASSAGE OF TYPHOON

Although our knowledge about the fate of phytoplankton bloom after the passage of typhoon is still rudimentary, two major fates are reported; 1) sinking into the deeper waters (e.g., Hung et al. 2010) and 2) grazed by meso- and microzooplankton (e.g., Chung et al. 2012).

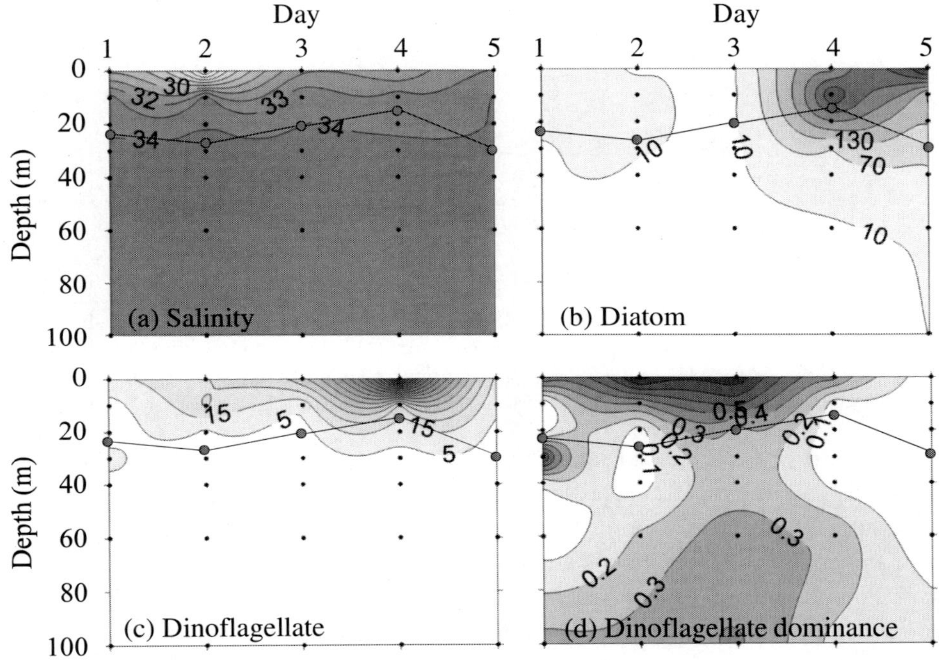

Modified from Tsuchiya 2014.

Figure 5. Temporal and vertical variations in (a) salinity, (b) diatom biomass (mg C m^{-3}), (c) dinoflagellate biomass (mg C m^{-3}) and (d) the ratio of dinoflagellate biomass to total phytoplankton biomass (dinoflagellate dominance) after the passage of *Malou* at the fixed offshore station in Sagami Bay, Japan. Solid lines with gray symbols indicate the euphotic depth (1%).

Regarding sinking, in the Philippine Sea, the dominant phytoplankton taxa changed from *Trichodesmium* spp. to diatoms after the typhoons (Chen et al. 2009). *Trichodesmium* spp. are positively buoyant and their biomass loss to sinking appears to be minor compared to cell lysis, extracellular release, and grazing, unlike diatoms (Mulholland 2007). Thus Chen et al. (2009) suggested that enhanced biogenic carbon produced attributed to diatoms is exported to deep waters than recycled in the shallow-water food web after the passage of typhoon. In Sagami Bay, a diatom bloom occurred 4 days after the passage of typhoon *Malou*, and then there was a relatively high diatom biomass even in deep waters below the euphotic layer, which suggests that increased diatoms in the euphotic zone might sink or be advected to deeper water (Figure 5; Tsuchiya 2014). In the southern East China Sea, using sediment trap before and after the passage of typhoon *Fengwong*, the particulate organic carbon (POC) flux 5 days after the passage of typhoon was 265±14 mg C m^{-2} d^{-1}, which was ~1.7-fold that

(140-180 m C m^{-2} d^{-1}) recorded during the period (June-August, 2007) when no typhoons occurred (Hung et al. 2010).

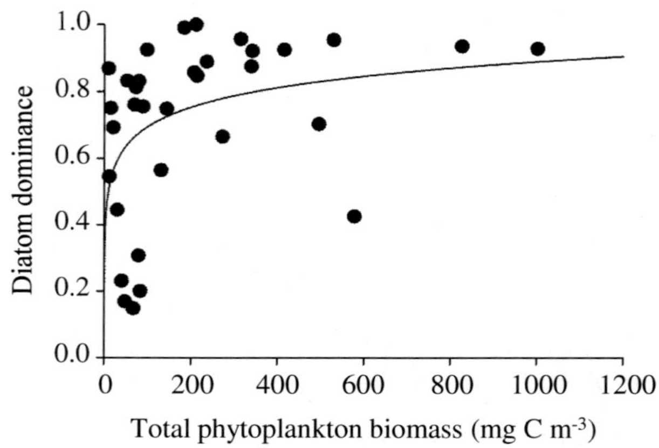

Modified from Tsuchiya 2014.

Figure 6. Relationship between diatom dominance (diatom biomass / total phytoplankton biomass) and total phytoplankton biomass [Diatom dominance = 0.31 + 0.084 × ln(total phytoplankton biomass); n = 33, r = 0.39, $P < 0.05$] after the passages of *Mawar*, *Sinlaku*, *Etau* and *Malou* at Manazuru Port in Sagami Bay, Japan.

And the ratio of POC flux to integrated primary production estimated by the model of Behrenfeld and Falkowski (1997) was in the range from 0.10 to 0.16 (Hung et al. 2010). As to grazing, Chung et al. (2012) observed a termination of diatom bloom and a concurrent increase in copepods and fecal pellets after the passage of typhoon *Morakot* in the subtropical Northwest Pacific, and suggested that intensive grazing pressure was the main cause of the termination of the diatom bloom. After the passage of typhoon *Fengshen* in the northeastern South China Sea, dilution experiments showed that the microzooplankton grazing rate significantly correlated with phytoplankton growth rate (Zhou et al. 2011). In southern Kaneohe Bay, Hawaii, a storm-runoff event caused a phytoplankton bloom, and a subsequent increase in copepods and appendicularians occurred (Hoover et al. 2006). In Sagami Bay, pheopigment concentrations increased to a peak 5 days after the passage of typhoon *Malou* accompanied by phytoplankton bloom (Tsuchiya 2014).

Pheopigments are degradation products of algal chlorophyll pigments, and are known to be direct products of zooplankton grazing (Shuman and Lorenzen 1975, Welschmeyer et al. 1984, Welschmeyer and Lorenzen 1985).

Moreover, zooplankton grazing becomes more active in a turbulent environment (Kiørboe 1993, Hwang et al. 1994).

Thus, the results suggest that zooplankton grazing pressure on phytoplankton might have intensified. However, it has not been quantified what proportion of new production is grazed after the passage of typhoon.

A large amount of fish larvae occurs in Sagami Bay and Tokyo Bay from summer to autumn, such as the Japanese anchovy *Engraulis japonicas*, Japanese sardine *Sardinops melanostictus*, *Sardinella zunasi*, *Sillago japonica* and Perciform species (Nakata and Mitani 1979, Kudo 1991, Nagaiwa et al. 2005, Sassa and Kawaguchi 2006).

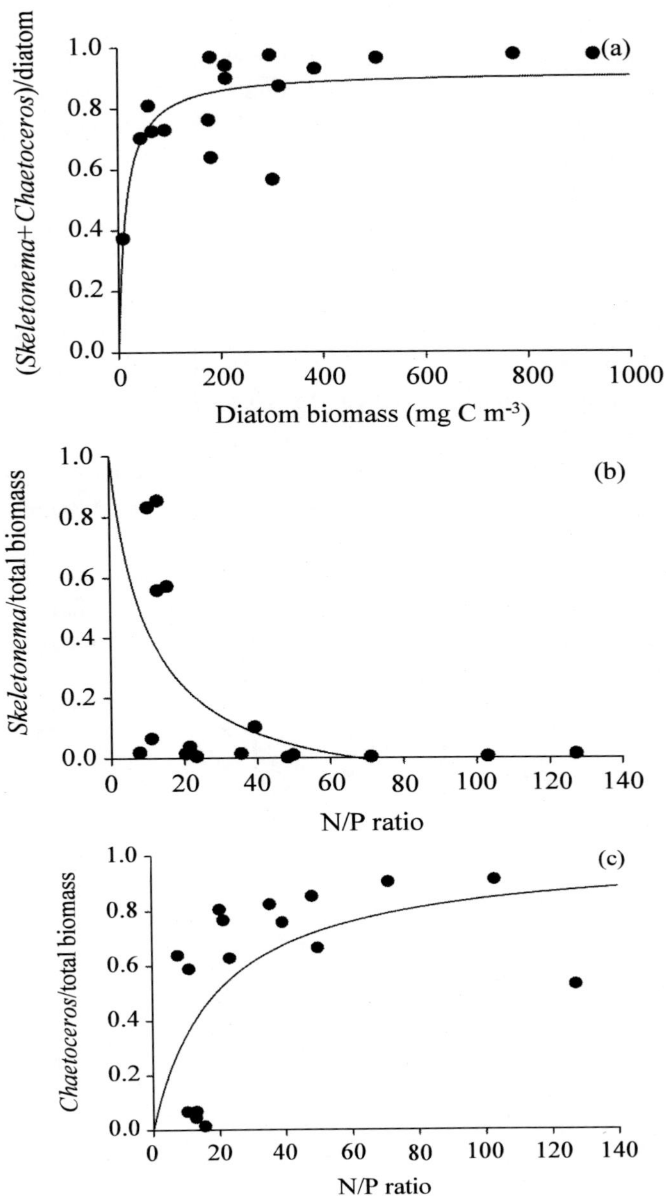

Modified from Tsuchiya et al. 2014.

Figure 7. Relationships between (a) proportion of *Skeletonema* spp. + *Chaetoceros* spp. in diatom biomass and diatom biomass, (b) proportion of *Skeletonema* spp. in total biomass and N/P ratio, and (c) proportion of *Chaetoceros* spp. in total biomass and N/P ratio after the passages of *Mawar*, *Sinlaku*, *Etau* and *Malou* at Manazuru Port in Sagami Bay, Japan. The regression curves were (a) *Skeletonema* + *Chaetoceros* / diatom biomass = 0.92 × diatom biomass / (15 + diatom biomass); $n = 17$, $r = 0.75$, $P < 0.05$, (b) *Skeletonema* / total biomass = 1 – 1.2 × N/P ratio / (10 + N/P ratio); $n = 16$, $r = -0.56$, $P < 0.05$) and (c) *Chaetoceros* / total biomass = 1.0 × N/P ratio / (19 + N/P ratio); $n = 16$, $r = 0.60$, $P < 0.05$). Data of (a) whose total dominance of *Skeletonema* spp. and *Chaetoceros* spp. in the diatom community exhibited more than 50% was used for analyses of (b) and (c).

Mortality occurs mostly during the pelagic larval stage, and the rate of larval mortality is directly reflected in the population dynamics of fishes (e.g., May 1974). After the absorption of the yolk sac, fish-larvae must obtain sufficient food to meet their metabolic requirements within 2-3 days or they will die (Smith and Lasker 1978), and the survival of first-feeding larvae is affected by food concentrations (e.g., Houde 1977, Lasker and Smith 1977). Feeding in anchovy larvae is stimulated when the phytoplankton concentration is more than > 20-30 cells mL^{-1} (Lasker 1975). The phytoplankton cell density after the passage of typhoons in Sagami Bay reached up to 1.5×10^4 cells mL^{-1} (Tsuchiya et al. 2013), which might be a favorable condition for first-feeding fish larvae and enhance their survival rate. Thus, the passage of typhoons might also contribute to the high fishery production in the bay.

Recent analyses of past typhoons (including tropical cyclones and hurricanes) have suggested that global warming leads to increasing intensities of wind and rainfall which in turn, causes episodic events (Emanuel 2005, Webster et al. 2005, Elsner et al. 2008, Knutson et al. 2010, Yamada et al. 2010). With the intensification of typhoons, diatom blooms and the associated carbon export are expected to increase in magnitude since loading of nutrients and organic matter could be augmented.

The higher production and more efficient carbon export to deeper water may result in increased CO_2 draw down from atmosphere through the biological pump. The process can lead to a negative feedback of the natural earth system on global warming (Figure 8), in other words, leading to global cooling as a result of organic carbon sedimentation.

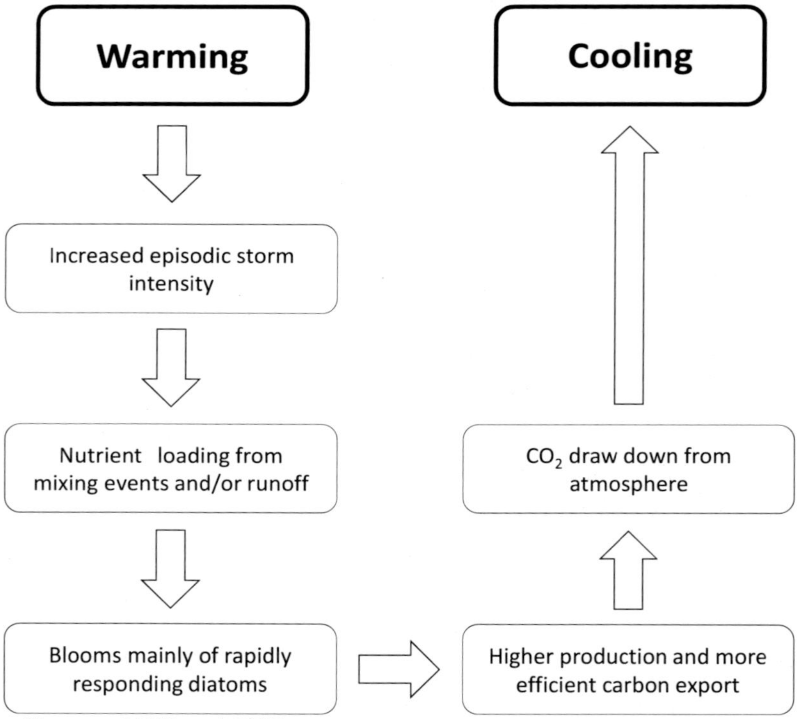

Modified from Kemp and Villareal 2013.

Figure 8. Potential negative feedback on global warming. The various paths show mechanisms for increasing and/or favouring increased diatom production leading to increased carbon export and consequent enhanced CO_2 draw down, promoting global cooling.

In future studies, the fate of biological production induced by typhoon should be investigated in order to better understand the effects of typhoon on ocean biogeochemical cycles. These studies would include objectives such as (1) to measure the primary production and POC export flux using a sediment trap before and after the passage of typhoon in order to assess how much typhoon-derived biogenic carbon is exported to deeper waters, and (2) to examine the grazing rates of zooplankton on primary producers before and after the passage of typhoon in order to assess the functional changes of the food web.

CONCLUSION

Typhoons are important disturbance events occurring in the tropical, subtropical and temperate regions (e.g., Tseng et al. 2010, Chung et al. 2012). Physical disturbances such as terrestrial runoff, sediment resuspension, vertical mixing and upwelling supply a large amount of nutrients to the euphotic zone, and enhance phytoplankton productivity significantly.

The enhanced primary production after the passage of typhoon could account for up to 9.1% of the annual primary production in the upper waters.

Meteorological disturbances such as typhoons, which will likely intensify as global climate change progresses, should be considered in annual production models. Phytoplankton blooms are dominated by diatoms, and the phytoplankton species succession in diatom communities could be related to nutrient stoichiometry, which is determined by relative contributions of various nutrient sources.

Many previous studies have revealed increase in phytoplankton biomass and production, and change of species composition after a typhoon event (e.g., Tsuchiya et al. 2013).

In the future, the fate of phytoplankton blooms after a typhoon passage should be examined in order to evaluate the contribution of typhoons to fishery resources and biogeochemical cycles, with the wider hope of eventually understanding the role of typhoons in supporting the earth's homeostatic processes.

REFERENCES

Ara, K., Hiromi, J. (2009) Seasonal variability in plankton food web structure and trophodynamics in the neritic area of Sagami Bay, Japan. *J. Oceanogr.* 65:757-779.

Behrenfeld, M. J., Falkowski, P. G. (1997) Photosynthetic rates derived from satellite-based chlorophyll concentration. *Limnol. Oceanogr.* 42:1-20.

Chang, J., Chung, C., Gong, G. (1996) Influences of cyclones on chlorophyll a concentration and Synechococcus abundance in a subtropical western Pacific coastal ecosystem. *Mar. Ecol. Prog. Ser.* 140:199-205.

Chen, Y. L., Chen, H., Jan, S., Tuo, S. (2009) Phytoplankton productivity enhancement and assemblage change in the upstream Kuroshio after typhoons. *Mar. Ecol. Prog. Ser.* 385:111-126.

Chung, C. C., Gong, G. C., Hung, C. C. (2012) Effect of Typhoon Morakot on microphytoplankton population dynamics in the subtropical Northwest Pacific. *Mar. Ecol. Prog. Ser.* 448:39-49.

Delesalle, B., Pichon, M., Frankignoulle, M., Gattuso, P. J. (1993) Effects of a cyclone on coral reel phytoplankton biomass, primary production and composition (Moorea Island, French Polynesia). *J. Plankton Res.* 15: 1413-1423.

Donaghay, P. L., Osborn, T. R. (1997) Toward a theory of biological-physical control of harmful algal bloom dynamics and impacts. *Limnol. Oceanogr.* 42:1283-1296.

Edmond, J. M., Spivack, A., Grant, B. C., Hu, M. H., Chen, X., Cheng, S., Zeng, X. (1985) Chemical dynamics of the Changjiang Estuary. *Cont. Shelf Res.* 4: 17-36.

Elsner, B. J., Kossin, P. J., Jagger, H. T. (2008) The increasing intensity of the strongest tropical cyclones. *Nature* 455:92-95.

Emanuel, K. (2005) Increasing destructiveness of tropical cyclones over the past 30 years. *Nature* 436:686-688.

Escaravage, V., Prins, T. C., Smaal, A. C., Peethers, J. C. H. (1996) The response of phytoplankton communities to phosphorus input reduction in mesocosm experiments. *J. Exp. Mar. Biol. Ecol.* 198:55-79.

Estrada, M., Berdalet, E., Vila, M., Marrase, C. (2003) Effects of pulsed nutrient enrichment on enclosed phytoplankton: ecophysiological and successional responses. *Aquat. Microb. Ecol.* 32:61-71.

Fogel, M. L., Aguilar, C., Cuhel, R., Hollander, D. J., Willey, J. D., Paerl, H. W. (1999) Biological and isotopic changes in coastal waters induced by Hurricane Gordon. *Limnol. Oceanogr.* 44:1359-1369.

Fujiki, T., Toda, T., Kikuchi, T., Aono, H., Taguchi, S. (2004) Phosphorus limitation of primary productivity during the spring-summer blooms in Sagami Bay, Japan. *Mar. Ecol. Prog. Ser.* 283:29-38.

Furnas, M. J. (1989) Cyclonic disturbance and a phytoplankton bloom in an tropical shelf ecosystem. In: Okaichi, T., Anderson, D. M., Nemoto, T. (eds.) *Red Tides: Biology, Environmental Science and Toxicology*. Elsevier, New York, p. 273-276.

Furnas, M. J. (1990) In situ growth rates of marine phytoplankton' approaches to measurement, community and species growth rates. *J. Plankton Res.* 12:1117-1151.

Furnas, M. J. (1991) Net *in situ* growth rates of phytoplankton in an oligotrophic, tropical shelf ecosystem. *Limnol. Oceanogr.* 36:3-29.

Glynn, P. W., Almodovar, L. R., Gonzalez, J. G. (1964) Effects of hurricane Edith on marine life in La Parguera, Puerto Rico. *Caribb. J. Sci.* 4:335-345.

Hansen, P. J. (1991) Quantitative importance and trophic role of heterotrophic dinoflagellates in a coastal pelagial food web. *Mar. Ecol. Prog. Ser.* 73: 253-261.

Ho, T. Y. (2006) The trace metal composition of marine microalgae in cultures and natural assemblages. In: Subba Rao, D. V. (ed.) *Algal Cultures, Analogues of Blooms and Applications,* Science Publishers, Enfield NH, p. 271-300.

Hoover, R. S., Hoover, D., Mitter, M., Landry, M. R., DeCarlo, E. H., Mackenzie, F. T. (2006) Zooplankton response to storm runoff in a tropical estuary: bottom-up and top-down controls. *Mar. Ecol. Prog. Ser.* 318:187-201.

Houde, E. D. (1977) Food concentration and stocking density effects on survival and growth of laboratory-reared larvae of bay anchovy *Anchoa mitchilli* and lined sole *Achirus lineatus*. *Mar. Ecol.* 43:333-341.

Hung, C. C., Gong, G. C., Chou, W. C., Chung, C. C., Lee, M. A., Chang, Y., Chen, H. Y., Huang, S. J., Yang, Y., Yang, W. R., Chung, W. C., Li, S. L., Laws, E. (2010) The effect of typhoon on particulate organic carbon flux in the southern East China Sea. *Biogeosciences* 7:3007-3018.

Hwang, J. S., Costell, J. H., Strickler, J. R. (1994) Copepod grazing in turbulent flow: elevated foraging behavior and habituation of escape responses. *J. Plankton Res.* 16:421-431.

Ismael, A. A. (2003) Succession of heterotrophic and mixotrophic dinoflagellates as well as autotrophic microplankton in the harbor of Alexandria, Egypt. *J. Plankton Res.* 25:193-202.

Jacobson, D. M. (1999) A brief history of dinoflagellate feeding research. *J. Eukaryot. Microbiol.* 46:376-381.

Jeong, H. J. (1999) The ecological roles of heterotrophic dinoflagellates in marine planktonic community. *J. Eukaryot. Microbiol.* 46:390-396.

Kamatani, A., Oku, O., Tsuji, H., Maeda, M., Yamada, Y. (2000) The distribution and fate of nutrients in Sagami Bay. *Nippon Suisan Gakkaishi* 66:70-79 (in Japanese).

Katz, M. E., Finkel, Z. V., Grzebyk, D., Knoll, A. H., Falkowski, P. G. (2004) Evolutionary trajectories and biogeochemical impacts of marine eukaryotic phytoplankton. *Annu. Rev. Ecol. Evol.* S 35:523-556.

Kemp, A. E. S., Villareal, T. A. (2013) High diatom production and export in stratified waters - A potential negative feedback to global warming. *Prog. Oceanogr.* 119:4-23.

Kiørboe, T. (1993) Turbulence, phytoplankton cell size and the structure of pelagic food webs. *Adv. Mar. Biol.* 29:1-72.

Knutson, T. R., McBride, J. L., Chan, J., Emanuel, K., Holland, G., Landsea, C., Held, I., Kossin, J. P., Srivastava, A. K., Sugi, M. (2010) Tropical cyclones and climate change. *Nature Geosci.* 3:157-163.

Kudo, T. (1991) Sudden increase of Japanese sardine larvae, *Sardinops melanostictus* (Temmnck et Schlegel), during the period from October to December in Sagami Bay. *Report of Kanagawa fisheries experimental station* 12:73-82 (in Japanese).

Lagus, A., Suomela, J., Weithoff, G., Heikkila, K., Helminen, H., Sipura, J. (2004) Species-specific differences in phytoplankton responses to N and P enrichments and the N:P ratio in the Archipelago Sea, northern Baltic Sea. *J. Plankton Res.* 26:779-798.

Lasker, R. (1975) Field criteria for the survival of anchovy larvae: the relation between inshore chlorophyll maximum layers and successful first feeding. *Fish Bull. US* 73:847-855.

Lasker, R., Smith, P. E. (1977) Estimation of the effects of environmental variations on the eggs and larvae of the northern anchovy. *CalCOFI Rep.* 19:128-137.

Lessard, E. J., Swift, E. (1985) Species-specific grazing rates of heterotrophic dinoflagellates in oceanic waters, measured with a dual-label radioisotope technique. *Mar. Biol.* 87:289-296.

Lin, I., Liu, W. T., Wu, C., Wong, G. T. F., Hu, C., Chen, Z., Liang, W., Yang, Y., Liu, K. (2003) New evidence for enhanced ocean primary production triggered by tropical cyclone. *Geophys. Res. Lett.* 30:1718, doi:10.1029/ 2003GL017141

Liu, K. K., Iseki, K., Chao, S. Y. (2000) Continental margin carbon fluxes. In: Hanson, R. B. (ed.) *The Changing Ocean Carbon Cycle, IGBP book series*, Cambridge University Press, p. 187-239.

Liu, K., Chao, S., Shaw, P., Gong, G., Chen, C., Tang, T. (2002) Monsoon-forced chlorophyll distribution and primary production in the South China Sea: observations and a numerical study. *Deep-Sea Res.* Part I 49:1387–1412.

Loftus, M. E., Seliger, H. H. (1976) A comparative study of primary production and standing crop of phytoplankton in a portion of the upper Chesapeake Bay subsequent to Tropical Storm Agnes. In: Davis, J. (ed.), *The effects of Tropical Storm Agnes on the Chesapeake Bay Estuarine System.* Chesapeake Res. Consortium Pub. no. 54. The Johns Hopkins University Press, Baltimore, p. 509-521.

May, R. C. (1974) Larval mortality in marine fishes and the critical period concept. In: Blaxter, J. H. S. (ed.) *The early life history of fish.* Springer-Verlag Berlin Heidelberg, New York, p. 3-19.

McKinnon, A. D., Meekan, M. G., Carleton, J., Furnas, M. J., Duggan, S., Skirving, W. (2003) Rapid changes in shelf waters and pelagic communities on the southern Northwest Shelf, Australia, following a tropical cyclone. *Cont. Shelf Res.* 23:93-111.

Miyaguchi, H., Fujiki, T., Kikuchi, T., Kuwahara, V. S., Toda, T. (2006) Relationship between the bloom of *Noctiluca scintillans* and environmental factors in the coastal waters of Sagami Bay, Japan. *J. Plankton Res.* 28:313-324.

Møller, M., Myklestad, S., Haug, A. (1975) Alkaline and acid phosphatases of the marine diatoms *Chaetoceros affinis* var. *willei* (Gran) Hustedt and *Skeletonema costatum* (Grev.) Cleve. *J. Exp. Mar. Biol. Ecol.* 19:217-226.

Morton, S. L. (2000) Phytoplankton ecology and distribution at Manatee Cay, Perican Cays, Belize. *Atoll. Res. B* 472:125-132.

Mulholland, M. R. (2007) The fate of nitrogen fixed by diazotrophs in the ocean. *Biogeosciences* 4:37-51.

Nagaiwa, R., Moteki, M., Nohno, H., Fujita, K. (2005) Larval and juvenile fish assemblages in surface waters at the mouth of Tokyo Bay. *La mer* 43: 97-104 (in Japanese).

Nakata, N., Mitani, I. (1979) Occurrence and distribution of eggs and larvae in Kaneda Bay, Kanagawa Prefecture. *Report of Kanagawa fisheries experimental station* 79-10:81-89 (in Japanese).

Nishitani, G., Miyamura, K., Ima, I. (2003) Trying to cultivation of *Dinophysis cautdata* (Dinophyceae) and the appearance of small cells. *Plankton Biol. Ecol.* 50:31-36.

Paerl, H. W., Bales, J. D., Ausley, L. W., Buzzelli, C. P., Crowder, L. B., Eby, L. A., Fear, J. M., Go, M., Peierls, B. L., Richardson, T. L., Ramus, J. S. (2001) Ecosystem impacts of three sequential hurricanes (Dennis, Floyd and Irene) on the Unites States' largest lagoonal estuary, Pamlico Sound, NC. *Proc. Natl. Acad. Sci. US* 98:5655-5660.

Paerl, H. W., Valdes, L. M., Adolf, J. E., Peierls, B. M., Harding, L. J. (2006) Anthropogenic and climatic influences on the eutrophication of large estuarine ecosystems. *Limnol. Oceanogr.* 51:448-462.

Price, J. F. (1981) Upper ocean response to a hurricane. *J. Phys. Oceanogr.* 11:153-175. Sverdrup, H. U., Johnson, M. W., Fleming, R. H. (1942) *The Oceans, Their Physics, Chemistry and Biology.* Prentice-Hall, Englewood Cliff, NJ, 1087 p.

Raven, J. A. (1997). The vacuole: a cost-benefit analysis. *Adv. Bot. Res.* 25: 59-86.

Reinfelder, J. R., Kraepiel, A. M., Morel, F. M. (2000) Unicellular C_4 photosynthesis in a marine diatom. *Nature* 407:996-999.

Rees, A. P., Joint, I., Donald, K. M. (1999) Early spring bloom phytoplankton-nutrient dynamics at the Celtic Sea Shelf Edge. *Deep Sea Res. II* 46:483-510.

Reinfelder, J. R., Milligan, A. J., Morel, F. M. (2004) The role of the C_4 pathway in carbon accumulation and fixation in a marine diatom. *Plant Physiol.* 135:2106-2111.

Roden, C. M., O'Mahony, K. W. (1984) Competition as a mechanism of adaptation to environmental stress in outdoor cultures of marine diatoms. *Mar. Ecol. Prog. Ser.* 16:219-227.

Ryther, J. H. (1955) Ecology of autotrophic marine dinoflagellates with reference to red water conditions. In: Johnson, F. H. (ed.) The luminescence of biological systems. *AAAS*, p. 387-413.

Sassa, C., Kawaguchi, K. (2006) Occerrence patterns of mesopelagic fish larvae in Sagami Bay, central Japan. *J. Oceanogr.* 62:143-153.

Satoh, F., Hamasaki, K., Toda, T., Taguchi, S. (2000) Summer phytoplankton bloom in Manazuru Harbor, Sagami Bay, central Japan. *Plankton Biol. Ecol.* 47:73-79.

Sellner, K. G., Sellner, S. G., Lacouture, R. V., Magnien, R. E. (2001) Excessive nutrients select for dinoflagellates in the stratified Patapsco River esturary: Margalef reigns. *Mar. Ecol. Prog. Ser.* 220:93-102.

Sherr, E. B., Sherr, B. F. (2007) Heterotrophic dinoflagellates: a significant component of microzooplankton and major grazers of diatoms in the sea. *Mar. Ecol. Prog. Ser.* 352:187-197.

Shiah, F. K., Chung, S. W., Kao, S. J., Gong, G. C., Liu, K. K. (2000) Biological and hydrographical responses to tropical cyclones (typhoons) in the continental shelf of the Taiwan Strait. *Cont. Shelf Res.* 20:2029-2044.

Shuman, F. R., Lorenzen, C. J. (1975) Quantitative degradation of chlorophyll by a marine hervivore. *Limnol. Oceanogr.* 20:580-586.

Smayda, T. J. (1997) Harmful algal blooms: Their ecophysiology and general relevance to phytoplankton blooms in the sea. *Limnol. Oceanogr.* 42: 1137-1153.

Smith, P. E., Lasker, R. (1978) Position of larval fish in an ecosystem. *Rapp. P-v Réun Cons. Int. Explor. Mer.* 173:77-84.

Strom, S. L., Morello, T. A. (1998) Comparative growth rates and yields of ciliates and heterotrophic dinoflagellates. *J. Plankton Res.* 20:571-584.

Sugawara, T., Hamasaki, K., Toda, T., Kikuchi, T., Taguchi, S. (2003) Response of natural phytoplankton assemblages to solar ultraviolet radiation (UV-B) in the coastal water, Japan. *Hydrobiologia* 493:17-26.

Sun, L., Yang, Y., Xian, T., Lu, Z., Fu, Y. (2010) Strong enhancement of chlorophyll a concentration by a weak typhoon. *Mar. Ecol. Prog. Ser.* 404:39-50.

Tilman, D., Kiesling, R., Sterner, R., Kilham, S. S., Johnson, F. A. (1986) Green, bluegreen and diatom algae: Taxonomic differences in competitive ability for phosphorus, silicon and nitrogen. *Arch. Hydrobiol.* 106:473-485.

Tozzi, S., Schofield, O., Falkowski, P. (2004) Historical climate change and ocean turbulence as selective agents for two key phytoplankton functional groups. *Mar. Ecol. Prog. Ser.* 274:123-132.

Tseng, Y. F., Hsu, T. C., Chen, Y. L., Kao, S. J., Wu, J. T., Lu, J. C., Lai, C. C., Kuo, H. Y., Lin, C. H., Yamamoto, Y., Xiao, T., Shiah, F. K. (2010) Typhoon effects on DOC dynamics in a phosphate-limited reservoir. *Aquat. Microb. Ecol.* 60:247-260.

Tsuchiya, K., Yoshiki, T., Nakajima, R., Miyaguchi, H., Kuwahara, V. S., Taguchi, S., Kikuchi, T., Toda, T. (2013) Typhoon-driven variations in primary production and phytoplankton assemblages in Sagami Bay, Japan: A case study of typhoon *Mawar* (T0511). *Plankton Benthos Res.* 8:74-87.

Tsuchiya, K. (2014) Ecological response of lower trophic levels to episodic typhoons in temperate coastal waters. *Doctor's thesis*, Soka University, Japan.

Tsuchiya, K., Kuwahara, V. S., Yoshiki, T., Nakajima, R., Miyaguchi, H., Kumekawa, N., Kikuchi, T., Toda, T. (2014) Phytoplankton community response and succession in relation to typhoon passages in the coastal waters of Japan. *J. Plankton Res.* 36:424-438.

Turner, R. E., Rabalais, N. N., Zhang, Z. N. (1990) Phytoplankton biomass, production and growth limitations on the Huanghe (Yellow River) continental shelf. *Cont. Shelf Res.* 10:545-571.

Valdes-Weaver, L. M., Piehler, M. F., Pinckney, J. L., Howe, K. E., Rossignol, K., Paerl, H. W. (2006) Long-term temporal and spatial trends in phytoplankton biomass and class-level taxonomic composition in the hydrologically variable Neuse-Pamlico estuarine continuum, North Carolina, US. *Limnol. Oceanogr.* 51:1410-1420.

Walker, N. D., Leben, R. R., Balasubramanian, S. (2005) Hurricane-forced upwelling and chlorophyll a enhancement within cold-core cyclones in the Gulf of Mexico. *Geophys. Res. Lett.* 32:L18610, doi:10.1029/2005GL 023716.

Webster, P. J., Holland, G. J., Curry, J. A., Chang, H. R. (2005) Changes in tropical cyclone number, duration, and intensity in a warming environment. *Science* 309:1844-1846.

Welschmeyer, N. A., Copping, A. E., Vernet, M., Lorenzen, C. J. (1984) Diel fluctuation in zooplankton grazing rate as determined from the downward vertical flux of pheopigments. *Mar. Biol.* 83:263-270.

Welschmeyer, N. A., Lorenzen, C. J. (1985) Chlorophyll budgets: Zooplankton grazing and phytoplankton growth in a temperate fjord and the Central Pacific Gyres. *Limnol. Oceanogr.* 30:1-21.

Wetz, M. S., Paerl, H. W. (2008) Estuarine phytoplankton responses to hurricanes and tropical storms with different characteristics (trajectory, rainfall, winds). *Estuar. Coast.* 31:419-429.

Yamada, Y., Oouchi, K., Satoh, M., Tomita, H., Yanase, W. (2010) Projection of changes in tropical cyclone activity and cloud height due to greenhouse warming: Global cloud-system-resolving approach. *Geophys. Res. Lett.* 37: L07709, doi:10.1029/2010GL 042518.

Zeeman, I. S. (1985) The effects of tropical storm Dennis on coastal phytoplankton. *Estuar. Coast. Shelf Sci.* 20:403-418.

Zhao, H., Tang, D., Wang, Y. (2008) Comparison of phytoplankton blooms triggered by two typhoons with different intensities and translation speeds in the South China Sea. *Mar. Ecol. Prog. Ser.* 365:57-65.

Zheng, M. G., Tang, D. (2007) Offshore and nearshore chlorophyll increases induced by typhoon winds and subsequent terrestrial rainwater runoff. *Mar. Ecol. Prog. Ser.* 333:61-74.

Zhou, L., Tan, Y., Huan, L., Huan, J., Liu, H., Lian, X. (2011) Phytoplankton growth and microzooplankton grazing in the continental shelf area of northeastern South China Sea after Typhoon Fengshen. *Cont. Shelf Res.* 31:1663-1671.

In: Natural Disasters
Editor: Karina Linwood

ISBN: 978-1-63463-309-3

Chapter 5

TYPHOON SWELL PREDICTION AND FREAK WAVE POTENTIAL

Nai-Kuang Liang*

Institute of Oceanography,
National Taiwan University, China

ABSTRACT

The typhoon or hurricane is an atmospheric eddy originated in tropical or subtropical ocean regions. Because the typhoon center always moves, then the wind field becomes non-uniform and non-stationary. Hence, the typhoon wave estimation is more complicated than that of the monsoon. However, the typhoon swell propagates outwards in all directions and reaches the shore earlier. As the typhoon is approaching a site, the swell wave height will be enhanced due to the accumulation of wave energy flux which bears some resemblance to Doppler effect. There are many cases that freak waves cause accidents and damages while typhoons are far away.

The assumption of the proposed typhoon swell prediction method is that the typhoon wave field is considered as a point source. The wave at the typhoon center is estimated by Bretschneider's parameter method. Empirical constants are obtained from measurements in Taiwan. The swell period will be longer, when the typhoon is more remote. The long wave is dangerous for the rock fishing, because the long wave can generate higher run-up and result to a larger backwash. However, the typhoon swell freak wave cannot be predicted but its potential can be alarmed.

INTRODUCTION

The typhoon or hurricane is an atmospheric eddy originated in tropical or subtropical ocean regions. Because the wind velocity turns rapidly, the wave generation process within typhoons becomes complicated. Parametric typhoon or hurricane wave prediction models were developed (Bretschneider & Tamaye, 1976; Young, 1987, Hsu, 1997), which pointed

* Email: liangnk@ntu.edu.tw

out that the maximum wave exists at the right side of typhoon center as one facing the forward direction. The typhoon or hurricane may generate huge waves within its domain. However, swells usually propagate faster than typhoon itself and appear outside the typhoon. Young (1987) stated also that SAR data from a number of hurricanes (King and Shemdin 1978; Gonzalez 1982; McLeish and Ross 1983; Beal et al. 1986) consistently showed swell ahead of hurricanes radiating out in a fan-shaped pattern from the center of the storm.

Freak waves are so-called rogue waves, which are considered nearly impossible according to traditional ocean wave theory and sometimes cause coastal and sea wrecks. There are already some hypotheses that explain freak wave occurrence and unfortunately we have many wave hazards around the Taiwan coast. These dangerous waves have been studied by Taiwanese ocean scientists and engineers and named as 'mad dog' waves, dangerous waves or freak waves (Tsai, Cheng-Han, Yin-Chern Lin and Hsiang-Mao Tseng, 2001; Chen, Guan-Yu, 2002; Chien, Hwa, Chia-Chuen Kao and Laurence Z. H. Chuang, 2002). Those waves can be categorized into four types (Dysthe, 2000). The first type may be regarded as extreme values of the statistical probability distribution tail. The second type is due to interaction between the waves and current. The third one is due to the wave-wave non-linear interaction. The fourth type is due to the superposition of waves of different frequency components. The first and fourth types belong to the statistical point of view, while the second and third types are physical aspects. The nonlinear process is modeled using a wave equation known as the nonlinear Schroedinger equation. In this equation a wave begins to 'soak' energy from the waves immediately fore and aft. In such a case, an unusual, unstable wave type may form a single wave that 'sucks' energy from other waves (Peregrine, D.H., 1985; Dole, J.W. and Peregrine, D.H., 1986, Henderson, K.L., Peregrine, D.H. and Dole, J.W., 1999). A simple statistical definition of the freak wave, which does not include steady current focusing, is a wave height that exceeds twice the significant wave height (Klinting and Sand, 1987; Kjeldsen, 1993).

There were three serious accidents taking place in Taiwan. (1) At about 2 p.m., October 23, 1987, regardless of the invasion of typhoon Lynn, 304 teachers and pupils of Hydraulic Elementary School of Pingtung County came to Mau-Bi-Tou coast (near southern tip of Taiwan) for a tour. As the pupils walked one by one along aisle to the coral reef, suddenly huge waves attacked the coast and 9 pupils were drowned in the sea. The whole nation dropped in a great grief. At the meantime, the typhoon scale has transferred from "strong" to "medium" and its center was about 500 km away but has moved fast previously toward Mau-Bi-Tou. On the next day, the headline of United Daily News wrote: "Typhoon Far Away, Hazard Close to Eyes" (Liang, 1990). (2) At about 4 a.m. on August 7, 1992, four fishing boats were totally destroyed by sudden huge waves in the vicinity of Suao Harbor at the east coast of Taiwan. One man died, two persons were missing and five fishermen were wounded. As the accident was close to the harbor, some wrecks were drifted to shore. One fisherman reminded that he has never confronted such big waves in his 40 years' fishing career'. Two days ago, a medium scale typhoon Janis had been in the area around 19°N, 136°E and moved fast toward Taiwan (Liang, 2006). (3) At about 2 p.m. on November 9, 2013, several huge long waves attacked the coast at the northeastern tip of Taiwan named Bitou Cape, which caused 8 deaths and 8 wounded tourists. During November 6 till 8, Super Typhoon-5 Haiyan moved fast in the sea area to the east of middle Philippines in WNW direction. The swell traveled more than 1600km to reach Bitou Cape and nobody acknowledged the danger, because there was no typhoon alarm.

Once, during the construction of Sen-Au Harbor in the northern Taiwan, an unfinished dolphin was destroyed by a typhoon swell, although the typhoon didn't invade Taiwan. This damage might be avoided, if a protection measure had been undertaken (Liang, 1990). Hualien Harbor facing the Pacific Ocean at the east coast of Taiwan suffers the harbor oscillation during the typhoon season. As a typhoon is still thousand kilometers away in the southeastern sector, the harbor oscillation may be so serious that ships must go out of the harbor to get rid of being damaged. A warning system, or a typhoon swell prediction scheme is useful to the harbor authority. In fact, such a prediction scheme is also helpful to our marine affair community. Typhoon Swell Doppler Effect

As a typhoon/hurricane approaches a location in the ocean quickly, the enhanced swell may be higher than those in the storm. If the typhoon approaching speed is close to the swell energy speed, i.e., the group velocity, a medium scale typhoon/hurricane may generate an extraordinary high swell. On Sept. 1, 2003, a medium scale typhoon Dujuan invaded southern Taiwan, which moved fast in WNW direction (Figure 1). Central Weather Bureau has a wave meter at Chengkung (Figure 1). The wave height data are shown in Figure 2. The measured peak wave height is close to 14 meters, which is larger than the estimated the wave inside the typhoon itself.

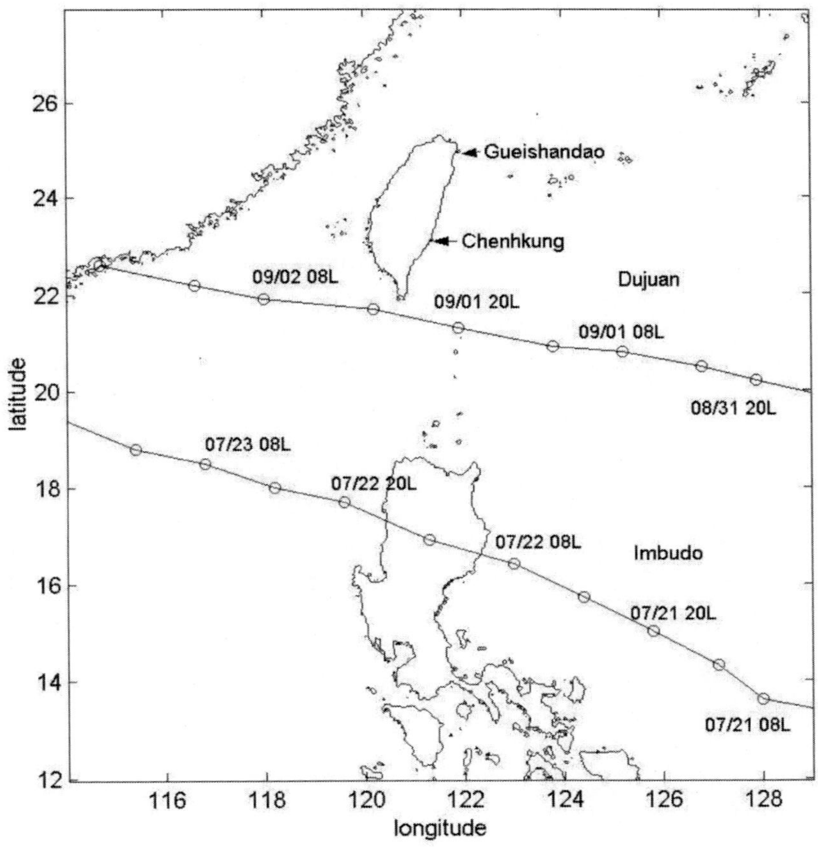

Figure 1. Typhoon tracks of Dujuan and Imbudo.

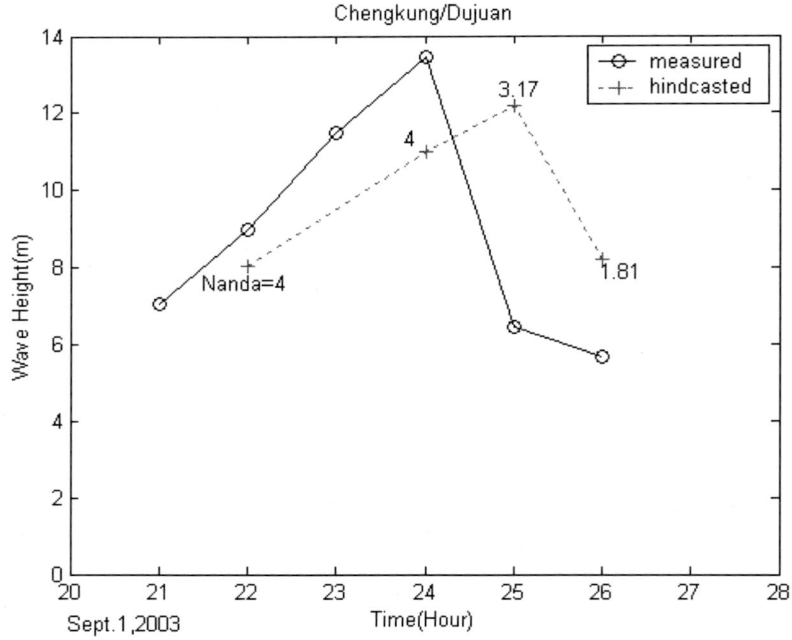

Figure 2. Measured wave height data at Chengkung during Dujuan typhoon.

The three above-mentioned accidents have the same common condition, i.e., a typhoon moving fast approaching toward the site, although it is very far away. The author named this wave height accumulation phenomenon as the typhoon swell Doppler effect (Liang, 2003). This can be explained as follows. The typhoon can be regarded as a wave generator. As the typhoon is stationary, the situation is different from that approaching a station. As the typhoon approaches the observer, the travel time for the wave energy becomes shorter. This results in an increase in the wave energy flux. This can be explained by the following: If Person A, who does not move, throws balls to Person B at a fixed time interval and the speed of the ball relative to the ground is constant, then Person B receives balls at the same time interval. If Person A moves toward Person B, Person B receives balls at a shorter time interval. If the ball is like the energy, Person B receives more energy flux from Person A, as Person A moves to Person B (Liang, 2006).

For engineering practice, the author proposed a swell height modification factor λ or called the "swell accumulation and relaxation" coefficient (Liang, 1982). If a typhoon is stationary, the swells appear within duration T_D. If the same typhoon is moving, the swell appearance time becomes T_D'. The total amount of energy passing through the cross section of a unit wave crest length is the same for the two cases (Liang, 2003).

$$\int_0^{T_D} (1/8)\rho g H^2 C_g dt = \int_0^{T_D'} (1/8)\rho g H'^2 C_g' dt \qquad (1)$$

in which ρ is the sea water density, g the gravitational acceleration and H or H' wave height for the two conditions respectively. For

$$C_g = C_g'$$

Then $\int_0^{T_D} H^2 dt = \int_0^{T_D'} H'^2 dt$

Assume that $H'/H = \lambda$, called the wave height modification factor and H or H' is not a function of time t. Then

$$\lambda = (T_D / T_D')^{1/2} \quad (2)$$

If T_1 and T_2 are the successive times for a typhoon and $Tlag_1$ and $Tlag_2$ are the corresponding time lags for the swell to arrive at the station. Then

$$T_D = T_2 - T_1 \quad (3)$$

and

$$T_D' = T_2 + Tlag_2 - (T_1 + Tlag_1) = T_D - (Tlag_1 - Tlag_2) = T_D - \Delta Tlag \quad (4)$$

where

$$\Delta Tlag = Tlag_1 - Tlag_2$$

$$Tlag = DD/(1.56T_S + V) \quad (5)$$

where V is the estimated ocean current component in the swell direction.

λ is greater than 1 as the typhoon approaches the station or its strength increases and less than 1 as the typhoon moves away from the station or its strength decreases. Due to the assumption of a stationary wave height in deriving equation (2), the time difference of two successive typhoon data should be less than 0.5 hour. Therefore, interpolated typhoon data and T_S data will be employed to estimate the successive lag times, i.e., T_{lag1} and T_{lag2}. As T_D' is very small, λ becomes unreasonable large. An upper limit for λ is set to be 4. While T_D' is negative, it means the latter wave overrun the former one. λ is set to be 1.41, i.e., an overlapping of two similar waves.

There was a perfect evidence for the typhoon swell Doppler effect (Liang, 2003). In August, 1995 typhoon Hent struck Taiwan. The typhoon center passed through the area (Figure 3 of the reference paper, Liang, 2003). A wave station was installed by Central

Weather Bureau (CWB) at Tungchi (Figure 3 of the reference paper, Liang, 2003), a small island in the Taiwan Strait. Both wave and wind data were available at that time, as shown in Figure 4 of the reference paper (Liang, 2003). From the same figure, the swell height rapidly increased from 1 m at 21 h, August 30 to 4.6m at 04 h, August 31. At 05 h, August 31 the wave height dropped suddenly to 2.78m. The wave period was maintained at about 10 s and the wind speeds were rather low until 02 h, August 31. The waves measured during this time were the typhoon swell. The low swell height at the initial stage was due to the sheltering effect of the southern part of Taiwan Island. The quick growth in the swell height and sudden drop is perfect evidence of the Doppler effect. The turning point in the high wave swells was between 04-05 h, August 31. From the geographical point of view, the turning point is when a typhoon changes from approaching to leaving. According to Figure 3 of the reference paper (Liang, 2003), the geographical turning point was the typhoon center at 23 h, August 30. The time lag was about 6 h, which matched the swell time propagation. The closed circle points in Figure 4 of the reference paper (Liang, 2003) are the measured swell heights.

TYPHOON SWELL PREDICTION SCHEME

Munk and Snodgrass (1957) discussed the swell height enhancement due to a moving storm. They proposed energy amplification μ which is a function of the wave spectrum, storm movement speed and storm and wave station positions, but without wind conditions. Dysthe and Harbitz (1987) studied big waves from polar lows, which have a rather limited horizontal extent of the order of 100 km but do not move slowly. As the movement speed is close to the wave group velocity, this phenomenon is called group velocity quasi-resonance. Donelan and Magnusson (2005) studied the probability of rogue wave occurrence due to mixed seas from two different wind systems, in which at least one of mixed sea components is swelling. They named this event meteorological focusing.

Following Munk and Snodgrass's goal, the author proposed a swell prediction scheme for a moving typhoon. The typhoon can be regarded as a wave generator. After Bretschneider's parametric hurricane wave model, wave heights within a hurricane are greater within a ring between R and 3R, in which R is the radius of the maximum wind speed, and decay as the one moves from the center outwards. Because wave energy is proportional to the square of the wave height, typhoon wave energy is concentrated at a ring between a radius of R and 3R. The typhoon can be regarded as a point source for the swell generation. Because the typhoon wind direction spreads to all directions, so do the waves. This phenomenon bears some resemblance to a pebble thrown into the water and generating ring waves. The wave height decays outwards from the center and is proportional to the reciprocal of the square root of the radius. The author assumed that the swell height at the ring of the typhoon radius of Beaufort scale No. 7 (R_7) is proportional to the wave height H_R at the radius of maximum wind speed (R). In this prediction scheme, H_R is estimated mainly by Bretschneider's parametric hurricane wave model (Bretschneider & Tamaye, 1976).

The employed typhoon/hurricane parameters are time, typhoon center location (latitude ϕ and longitude θ), central atmospheric pressure P_C in hPa and typhoon radius of Beaufort scale No. 7 (R_7) in km. The procedures are the following: Two semi-empirical equations were verified several times. The final form is as follows:

$$H_S = \lambda C_1 H_R \sqrt{\frac{R_7}{DD}} \tag{6}$$

$$T_S = C_2 T_P \tag{7}$$

in which H_S is the swell height in meters, λ swell height modification factor, C_1 and C_2 are empirical constants, H_R the estimated wave height in feet at the typhoon's radius of maximum wind speed R, R_7 the typhoon's radius of Beaufort's scale No. 7 wind in km, DD the distance between the typhoon center and the wave station in nautical mile, T_S the swell period, T_P the estimated wave peak period at R. DD is calculated by the following equations:

$$DD = 6371(-\tan^{-1}(Y/\sqrt{-Y^2+1}) + \pi/2)/1.852 \tag{8}$$

where $Y = \cos(\pi(90-\phi_0)/180)\cos(\pi(90-\phi)/180) +$

$\sin(\pi(90-\phi_0)/180)\sin(\pi(90-\phi)/180)\cos(\pi|\vartheta_0 - \vartheta|/180)$

ϕ_0 and θ_0 are the latitude and longitude of a wave meter or any interested location.

There are three parametric typhoon/hurricane wave forecasting methods, i.e., the methods proposed by Bretschneider and Tamaye (1976), Young (1987) and Hsu (1997). Hsu's method is the simplest. He uses only the maximum wind speed to determine wave heights. Young needed the maximum wind speed U_R, the radius of maximum wind speed R and the hurricane moving speed V_F. Bretschneider and Tamaye employed the central atmospheric pressure, the radius of the maximum wind speed R and the hurricane moving speed V_F. Bretschneider and Tamaye's method is more complicated and the results were proven to be reasonable. The author uses Bretschneider and Tamaye's method to estimate H_R in feet and T_P in seconds without considering the hurricane movement speed as in the following equation.

$$H_R = K'\sqrt{RdP} \tag{9}$$

$$T_P = 0.347\tanh\left\{\ln\left[\frac{1+\dfrac{40H_R}{(0.865U_R)^2}}{1-\dfrac{40H_R}{(0.865U_R)^2}}\right]^{0.5}\right\}^{0.6} \times U_R \tag{10}$$

where dP is the atmospheric pressure reduction from normal in inches of mercury and equal to the equation:

$$dP = (1000 + (1000 - P_C)/10 - P_C)/33.87 \tag{11}$$

where P_C is the typhoon central atmospheric pressure in hPa.

K' is a coefficient proposed by Bretschneider and Tamaye and a regression equation is as follows.

$$K' = 7.59 - 41.21(\frac{fR}{U_R}) + 160.51(\frac{fR}{U_R})^2 - 219.32(\frac{fR}{U_R})^3$$

Because the maximum wind speed radius R is not provided by the weather authority, a 'trial and error' method is used to estimate it by inputting the radius of Beaufort's scale No. 7 wind (R_7) and the maximum wind speed into the equation:

$$\frac{U_r}{U_R} = -\frac{1}{2}\frac{fR}{U_R}\frac{R_7}{R} + ((1+\frac{fR}{U_R})\frac{R}{R_7}\exp(1-\frac{R}{R_7}) + (\frac{1}{2}\frac{fR}{U_R}\frac{R_7}{R}))^{\frac{1}{2}} \tag{12}$$

where U_R is the geostrophic wind speed in knots as follows:

$$U_R = K\sqrt{dP} - 0.5fR \tag{13}$$

where f is the Coriolis parameter and equal to $0.5236\sin(\pi\phi/180)$ and the coefficient $K = -\phi/7.5 + 70$.

C_1 and C_2 are empirical constants, in which C_1 is then 0.106 and C_2 is a regression equation from 197 data (Liang, Lin and Tseng, 2010). The empirical equation is the following (Figure 3).

$$C_2 = 0.9393 + 0.0334\frac{DD}{T_P^{\,2}} \tag{14}$$

Because $T_P^{\,2}$ has a measure of wave length, $\dfrac{DD}{T_p^{\,2}}$ is an equivalent relative distance from a station to a typhoon center. It is obvious that C_2 increases as the equivalent relative distance increases.

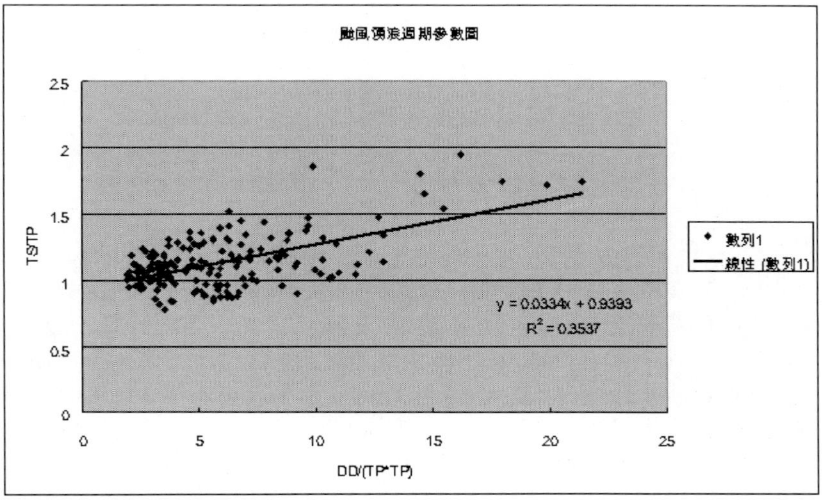

Figure 3. Regression line for C_2.

THE FREAK WAVE POTENTIAL OF TYPHOON SWELL

As ΔT_{lag} approaches T_D , i.e., the latter swell has almost caught the former swell, then $T_D^{'}$ approaches zero. Following equation (2), the wave height modification factor λ will be infinite. Of course it should not be infinite in reality but large. The definition of a freak wave is that the ratio of $H_{\max}$ to $H_{1/3}$ is larger than 2. Therefore, as the significant wave height $H_{1/3}$ is larger, then according to the 1st , 3rd and 4th type freak wave, a destructive freak wave is probable. Like the 2nd type freak wave λ makes the significant wave height larger. This makes sense! Other than typhoons or hurricanes, there are other wind systems that always move, such as extra-tropical cyclone, polar lows, etc. All these wind systems may have the same potential to generate freak waves.

As $T_D^{'}$ is smaller than T_D, λ becomes greater than one. As the distance between the typhoon center and wave station decreases and as C_g, i.e., a function of wave period, increases, $T_D^{'}$ would be smaller than T_D. Because λ is very sensitive to $T_D^{'}$, as the latter is small. As ΔT_{lag} is larger than T_D , i.e., the latter swell has overrun the former swell, then $T_D^{'}$ is negative and equation (2) is not applicable. Usually typhoon data are scattered. Any little typhoon data perturbation, such as positions of the typhoon center, central pressures, the radius of Beaufort scale No. 7 wind (R_7) , ocean current, etc., may change $T_D^{'}$. As $T_D^{'}$ is small, any change in the wind system may change its sign. Hence the sign of $T_D^{'}$ is not significant, an absolute value is employed. As the absolute value of $T_D^{'}$ is less than a small fraction of T_D , say one fourth, there is a high probability to have a freak wave. Freak Wave Potential Index Φ is defined as follows(Liang, 2012):

$$\Phi = \sqrt{\frac{T_D}{\left|T_D^{'}\right|}} \qquad (15)$$

The index Φ can be predicted. Once it is larger than 2 for any site at a time, a freak wave is highly probable. For the above-mentioned accident at Suao harbor in 1992, two days before, a medium scale typhoon Janis had been in the area around 19°N, 136°E and moved fast toward Taiwan. The typhoon Janis' track is shown in Figure 4. From 14h to 20h, Aug. 5, the typhoon central pressure decreased quickly from 970 hPa to 955 hPa. This made the latter wave period larger than the former one. The estimated "Freak Wave Potential Index" is 3.8! There was unfortunately no wave measurement. The source was at 20h, Aug. 5 and 790 nautical miles away, i.e., the 5th typhoon center from right in Figure 4. This was 32 hours before the accident.

Another freak wave potential criterion for coastal hazards is that the predicted swell height is larger than 2 meters, because the freak wave height is 2 times the significant wave height, i.e., larger than 4 meters. For a wave of long wave period, say greater than 10 seconds, the wave run-up and back wash is very strong on a rocky shore.

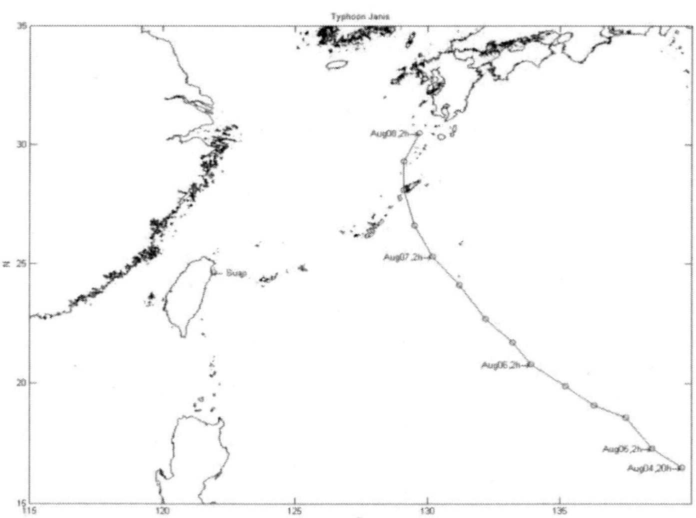

Figure 4. Track of Typhoon Jenis in 1992.

CONCLUSION

The index Φ is defined as the square root of a ratio, which is the swell appearance time for a stationary wind system divided by that for a moving and changing strength one, where an absolute value is taken for the denominator. As the denominator is positive, i.e., the later swell does not overrun the former, the index Φ reveals a wave height amplification coefficient. As the swell appearance time for a moving wind system (the denominator) $T_D^{'}$ is small, the index Φ is sensitive to the former, which is a function of the wind system strength and its approaching speed to the site. Usually the actual wind system data fluctuates around their mean values. Hence, the index Φ can be a measure of freak wave potential. As the "Wave Potential Index" is larger than 2, a freak wave is highly probable. This index can be correctly calculated because the historical typhoon datum is available. Furthermore, "Freak Wave Potential Index" concept may also be used for other moving wind systems.

Another freak wave potential criterion for coastal hazards is that the predicted swell height is larger than 2 meters and a freak wave height may be larger than 4 meters. For a wave of long wave period, the wave run-up and back wash are tremendous on a rocky shore. The proposed typhoon swell prediction scheme can serve the warning purpose.

ACKNOWLEDGMENT

The author thanks the Central Weather Bureau, Taiwan, Republic of China for the typhoon and wave data.

REFERENCES

Beal, R.C. et al. (1986) "Spacial variations of ocean wave directional spectra from the SEASAT synthetic aperture radar", J. Geophys. Res., 91(C2), 2433-2449.

Bretschneider, C.L. & Tamaye, E.E. (1976) "Hurricane wind and wave forecasting techniques", Proceedings 15th International Coastal Engineering Conference, 202-237.

Chen, Guan-Yu, (2002) "Possible explanations for coastal giant waves", *Journal of Coastal and Ocean Engineering*, Vol. 2, No. 1, pp.93-106

Chien, Hwa, Chia-Chuen Kao and Laurence Z. H. Chuang., (2002) "On the characteristics of observed coastal freak waves", *Coastal Engineering Journal*, Vol. 44, No. 4, pp.301-319.

Dole, J.W. and Peregrine, D.H., (1986) "Wave-wave modulation", *International Conference on Coastal Engineering,* ASCE, pp.163-175.

Donelan, M.A., A.K. Magnusson, (2005) "The role of meteorological focusing in generating rogue wave conditions", *'Aha Huliko'a Hawaiian Winter Workshop*, 7 p.

Dysthe, K.B. and A. Harbitz, (1987) "Big waves from polar lows?", Tellus, Vol. 39A, pp.500-508.

Dysthe, K.B., (2000) "Modeling a rogue wave — Speculations or realistic possibility?", *Abstract for Rogue Waves 2000 Workshop*, Brest, pp.29-30.

Gonzalez, F.I. (1982) "SEASAT wind and wave observations of northeast Pacific Hurricane Iva, August 13, 1978", J. Geophys. Res., 78(C5), 3431-3438.

Henderson, K.L., Peregrine, D.H. and Dole, J.W., (1999) "Unsteady water wave modulation: fully nonlinear solutions and comparison with the nonlinear Schroedinger equation", *Wave Motion*, Vol. 29, pp.341-361.

Hsu, S.A., (1997) "Forecasting maximum significant wave height induced by minimum central pressure of tropical storm at sea", Proc., *1997 CWB's Conference on Weather Analysis and Froecasting*, pp. 535-539.

King, D.B. & Shemdin, O.H. (1978) "Radar observations of hurricane wave directions", *Proc., 16th Int. Conf. On Coastal Engineering, Hamburg, West Germany,* 209-226.

Kjeldsen, S.P., (1993) "The wave follower experiment", *Proc. Symp. Air-Sea Interface Radio Acoustic Sensing Turbulence and Wave Dynamics*, Marseilles, France.

Klinting, P and Sand, S., (1987) "Analysis of prototype freak waves", *Coastal Hydrodynamics*, ASCE, pp. 618-632.

Liang, N.K. (1982) "A typhoon swell forecasting method", *Proceedings, the 6th Conference on Ocean Engineering*, Taiwan, R.O.C., 5-1～19 (in Chinese).

Liang, N.K. (1990) "A study on typhoon swell height prediction", ACTA OCEANOGRAPHICA TAIWANICA, No.25, 77-86.

Liang, N.K. (2003) "The typhoon swell Doppler effect", Ocean Engineering 30 1107-1115.

Liang, N.K. (2006) "The freak wave mystery—a new hypothesis for its occurrence", *Journal of Marine Science and Technology,* Vol. 15, No. 3, pp.241-246.

Liang, Nai-Kuang, Jiunn-Jye Lin and Hsiang-Mao Tseng,(2010) "A study on typhoon swell forecasting scheme", *Journal of Coastal Engineering*, Vol. 10, No. 2(2010), pp.220-236 (in Chinese).

Liang, Nai-Kuang (2012) "The freak wave potential of typhoon swell", *Journal of Marine Science and Technology,* Vol. 20, No. 4, pp.467-471.

McLeish, W. & Ross, D.B. (1983) "Imaging radar observations of directional properties of ocean waves" *J. Geophys. Res.*, 88(C7), 4407-4419.

Munk, W.H. and F.E. Snodgrass, (1957) "Measurements of southern swell at Guadalupe Island", *Deep Sea Research*, Vol. 4, pp. 272-286.

Peregrine, D.H., (1985) "Water wave and their development in space and time", *Proceedings of the Royal Society*, London, Vol. A400, pp. 1-18.

Tsai, Cheng-Han, Yin-Chern Lin and Hsiang-Mao Tseng, (2001) "Group waves and dangerous waves at the coast", *Journal of Coastal and Ocean Engineering*, Vol. 1, No. 1, pp.71-82.

Tulin, M.P., Y. Yao and A.K. Magnusson, (1996) "The evolution and structure of energetic wind waves", *Proc. 6th Int. Offshore and Polar Eng. Conf.*, Los Angeles, U.S.A, Vol. III, pp.1-17.

Young, Ian R. (1987) "Parametric hurricane wave prediction model", *Journal of Waterway, Port, Coastal and Ocean Engineering*, Vol. 114, No. 5, 637-652.

In: Natural Disasters
Editor: Karina Linwood

ISBN: 978-1-63463-309-3

Chapter 6

A TYPHOON DISASTER DATABASE SYSTEM

Takahiro Kayahara*[*], *Shinya Shimokawa* and *Ryohei Misumi
National Research Institute for Earth Science and Disaster Prevention, Japan

ABSTRACT

A typhoon disaster database system incorporating information about typhoon track, weather, and typhoon-related disasters, which have occurred in Japan since 1951, was developed and opened to the general public, via the Internet, beginning in July 2005. To construct the system, information was collected mainly from the Japan Meteorological Agency, the Fire and Disaster Management Agency, and local governments. The system shows the spatial distribution of typhoon disasters on a map using WebGIS. In this chapter, the function of the system is introduced, and the relationship between the number of users accessing the system and the characteristics of typhoons from 2006 to 2012 are discussed. The peak number of unique visitors each day was independent of typhoon intensity. However, it increased when typhoons passed through the Tokai and Kanto regions, where the population is relatively high.

Keywords: Typhoon disaster, Database, WebGIS, Peak unique visitors, Typhoon track

1. INTRODUCTION

Information on past typhoon disasters is very important for preparation against forthcoming typhoons. The Fifth World Meteorological Organization (WMO) International Workshop on Tropical Cyclones (IWTC-V) made the following recommendation to research communities: "An international database should be developed that would track the social and economic impacts of tropical cyclones as well as the costs associated with tropical cyclone forecasting, information services, and mitigation methods. This information could then be

[*] E-mail address: kayahara@bosai.go.jp

used to justify additional resources for tropical cyclone forecasting and disaster mitigation." [1].

Information on the damage caused by typhoon disasters was collected by local governments and governmental organizations in Japan and summarized as disaster reports. Recently, disaster reports have been made available on websites provided by the local governments and governmental organizations. These reports are dispersed as each organization issues their report independently, as such, collection of the disaster reports in Japan is a very laborious process.

The development of information and communication technology enables the provision of massive amounts of information through the Internet. This has been the case for typhoon information on the Internet. For example, the Japan Meteorological Agency (JMA) provides information on the number and tracks of typhoons and major typhoon disaster reports on their website [2, 3]. The National Institute of Informatics (NII) developed a digital typhoon site (http://agora.ex.nii.ac.jp/digital-typhoon/). It provides geostationary meteorological satellite images since 1978 and contains a weather disaster database extracted from weather anomalies and disasters data, provided by the JMA, and a typhoon damage database, extracted from Chronological Science Tables published by Maruzen [4]. The disaster information of these disaster databases is represented in text form.

However, we have developed a database system that stores data on typhoons and associated disasters in Japan since 1951 [5]. The system not only displays text information on the damage occurred, but also the damage distribution map on a web browser using WebGIS (Web Geographic Information System). The system also provides files to display the damage distribution on Google Earth.

In this chapter, we introduce the Typhoon Disaster Database System and describe the characteristics of the number of users accessing the system.

2. Typhoon Disaster Database System

2.1. Outline of Information Provided by the System

The system provides information on typhoons and related disasters that occurred in Japan after 1951. A tropical storm with 10-minute averaged maximum sustained winds of greater than 34 knots in the western North Pacific is called a typhoon in Japan. The typhoon number, name, GLIDE (Global unique disaster IDEntifier number, [6]), outline, track, intensity, track map, sea surface pressure map, infrared (IR) satellite image map, and precipitation map are stored in the system.

The typhoon number is provided by the JMA. It combines the last two digits of the calendar year and the two-digit serial number of the storm (e.g., T1411). The typhoon name used in the Joint Typhoon Warning Center (JTWC) (http://www.usno.navy.mil/JTWC/) was registered until 1999. Since 2000, the name adopted by the United Nations Economic and Social Commission for Asia and the Pacific (UNESCAP)/World Meteorological Organization (WMO) typhoon committee for the western North Pacific and the South China Sea, is registered. GLIDE is a globally common unique ID code for disasters, proposed by the Asian Disaster Reduction Center (ADRC) [6]. The outline of the typhoon situation is described based on the Geophysical Review by the JMA [7] or media (newspaper) information. The typhoon center position, central pressure, and maximum sustained wind speed are obtained from the best track data provided by the Regional Specialized Meteorological Center (RSMC) Tokyo–Typhoon Center (http://www.jma.go.jp/jma/jma-eng/jma-center/rsmc-hp-pub-eg/RSMC_HP.htm). The typhoon track maps are created using center position information in the best track data. The sea surface pressure maps are generated using reanalysis data provided by the National Centers for Environmental Prediction/The National Center for Atmospheric Research (NCEP/NCAR) [8]. The IR satellite image maps are created using IR satellite images for the area 70°N–20°S, 70°E–160°E distributed by Kochi University (http://weather.is.kochi-u.ac.jp/). The precipitation maps are created using hourly precipitation data from Automated Meteorological Data Acquisition System (AMeDAS) data.

Damage information related to typhoons has been primarily collected from the Geophysical Review by the JMA from 1951 to 2002 [7], damage reports by the Fire and Disaster Management Agency (FDMA) from 1999 [9], local governments, and media reports. There are differences in the number of damage incidents reported among information sources [10]. The content of damage information includes address, date, and damage situation. The damage situation is classified as shown in Table 1. The numbers and amounts of damages are registered in each item. In addition, a list of damage reports, which were used to extract damage information, a list of uniform resource locators (URLs), which relate typhoons and typhoon disasters, and a list of URLs for the disaster management offices of local governments, are registered.

Table 1. Types of damage and items

Type of damage	Items
Human damage	Dead or Missing and Injured
House damage	Ruined, Half ruined, Partly ruined, Inundation above floor level, Inundation below floor level, and Non-residential buildings
Agricultural damage	Farmlands, Agricultural facilities, Gardening facilities, Livestock, Livestock facilities, Crops, and others
Forest damage	Forest roads, Forest land, and others
Fisheries damage	Fishing boat, Fishing port, Fish firm facilities, and others
Public construction works damage	Rivers, Roads, Bridges, Sand arrestations, Harbors, Drains, Park, and others
Education facility damage	Schools, Kindergartens, Nursery schools, Social education facilities, Sports facilities, and others
Cultural property, sightseeing facility damage	Shrines and Temples, Sightseeing facilities, and others
Medical and Health facility damage	Medical facilities, Health facilities, and others
Commerce damage	Industries, Businesses, and others
Lifeline etc. damage	Water supply facilities, Electric power facilities, Gas supply facilities, Communication facilities, Railways, and others
Others	

2.2. Outline of the Display Information

The system contains the following main elements: a typhoon search page, a typhoon outline page, a page listing the damage situation, and a damage display page. Typhoons can be searched in various ways on the typhoon search page (Figure 1), i.e., by the typhoon number provided by the JMA, typhoon name, GLIDE, date, the prefecture and municipality where the typhoon damages occurred, and the location where the typhoon passed on a map. Figure 1 shows the typhoons tracks that caused damage in Japan in 2012.

Figure 1. An example of a typhoon search page that shows typhoons that caused damage in Japan in 2012. The blue, orange, red, and black lines indicate a tropical depression, typhoon, typhoon with maximum sustained wind speed of more than 64 knots, and extratropical cyclone, respectively.

The typhoon outline page contains an outline of the selected typhoon, the track map, sea surface pressure map, IR satellite image map, precipitation map, central pressure time series, and maximum sustained wind speed time series (Figure 2). The typhoon damage list page shows a list of the damage information presented by address. Detailed information of the damage appears in the list (Figure 3). The list can be narrowed by year, typhoon number, prefecture, information source, and type of damage, and sorted by year, typhoon number, date of typhoon generation, prefecture, municipality, and information source (Figure 3).

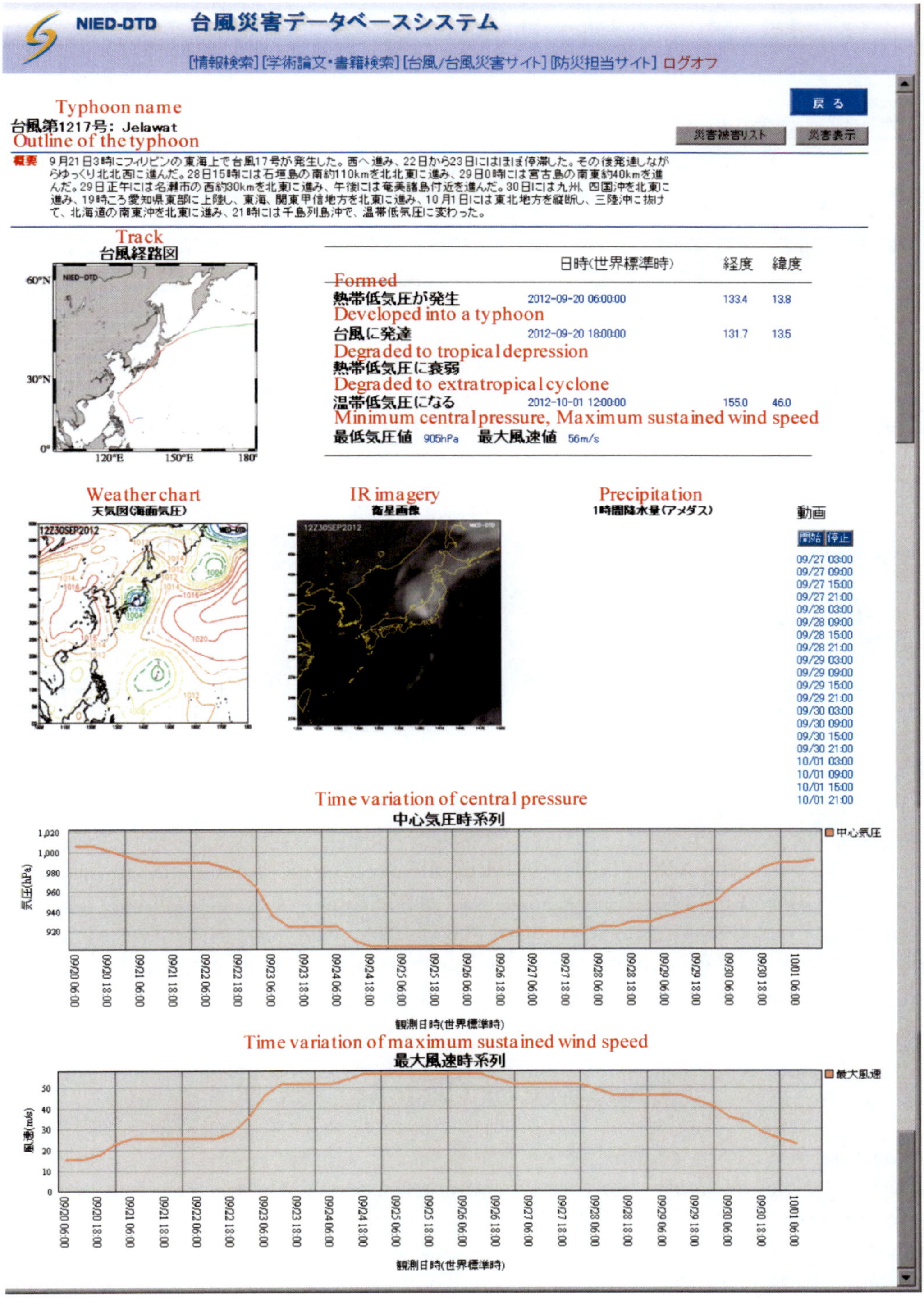

Figure 2. The typhoon outline page for typhoon 1217 (JELAWAT).

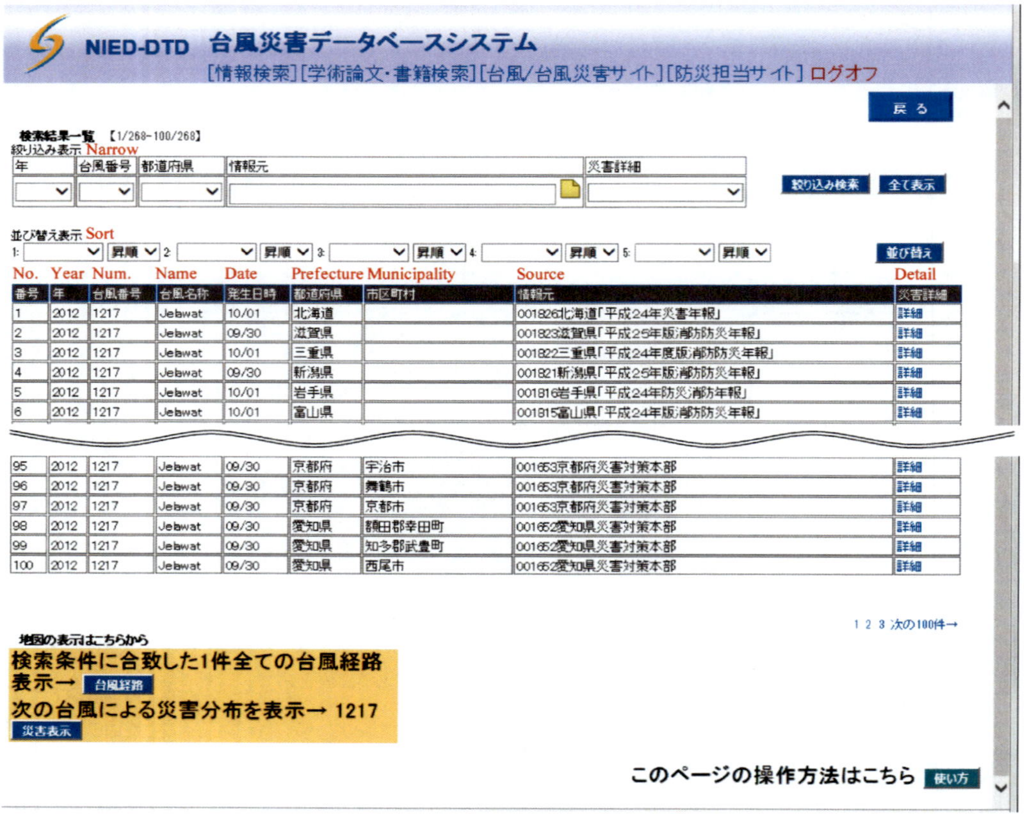

Figure 3. The typhoon damage list page for typhoon 1217 (JELAWAT).

The damage display page shows typhoon tracks and spatial distribution of the damage on a map using the WebGIS. When up to three types of damage are selected, the prefectures or municipalities where the selected damages occurred appear colored on the map (Figure 4). If the damage data are more precise than the municipality, the latitude and longitude information is additionally registered with the damage data, and a damage symbol is displayed on the map. Furthermore, the system can create a keyhole markup language (KML) file for Google Earth, which contains typhoon track and damage data (Figure 5).

3. Number of User Access

The system contains information on past typhoons and the related damage in Japan; therefore, it enables the user to learn about the disaster history in the place he/she lives.

To understand the current usage status of the system, the number of unique visitors was investigated from log files of the web server. Unique visitors were counted only once no matter how many times they revisited the web site within a specified period. A web log analyzer program "Visitors" was used to obtain the number of unique visitors daily from web log files (http://www.hping.org/visitors/).

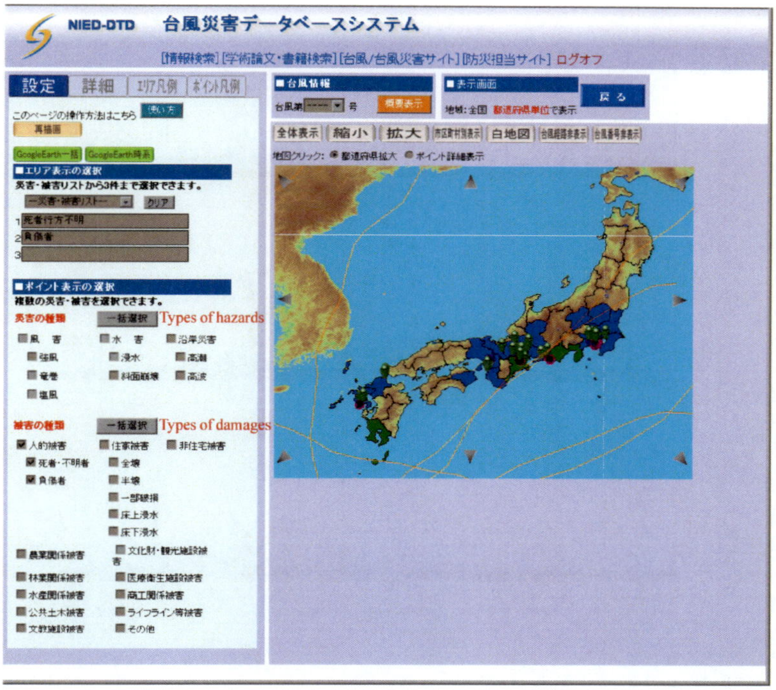

Figure 4. Typhoon damage display page, which shows the typhoon tracks for 2012, and the dead/missing and injured caused by 2012 typhoons. The green and blue prefectures indicate the injured and dead/missing and injured, respectively. The green and purple symbols indicate the injured and the occurrence of two or more damages.

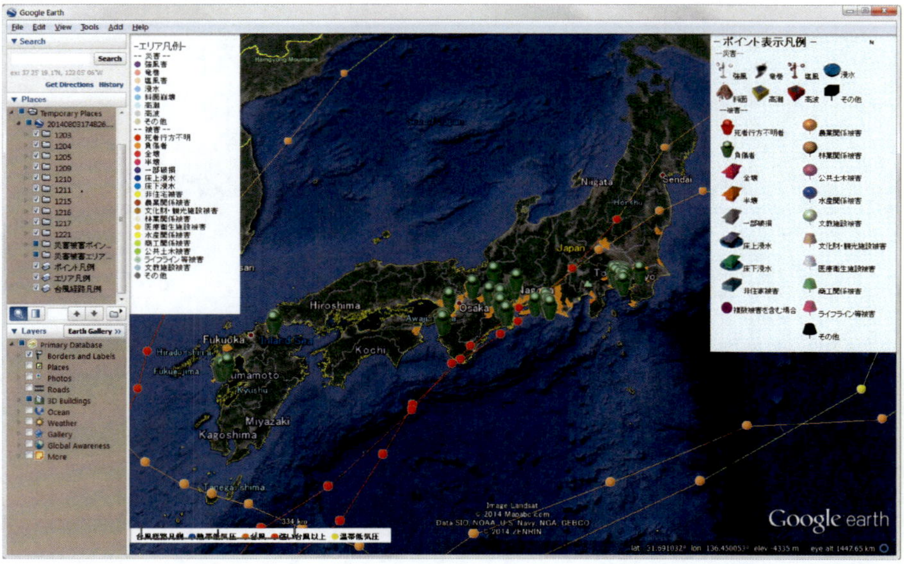

Figure 5. Example of Google Earth, showing the typhoon tracks of 2012, and injured caused. The orange region indicates the municipality where injuries occurred. The green symbol indicates the injured.

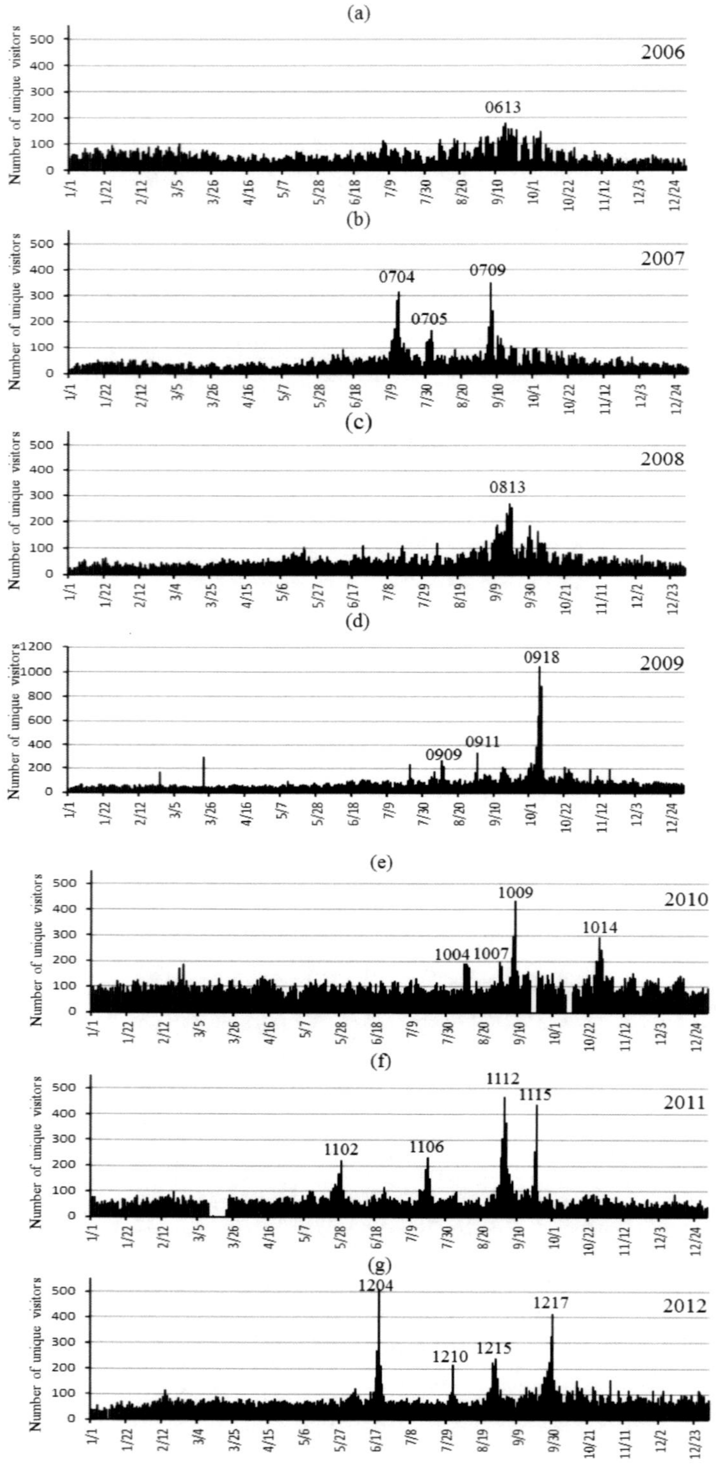

Figure 6. Time series of the daily number of unique visitors in (a) 2006, (b) 2007, (c) 2008, (d) 2009, (e) 2010, (f) 2011, and (g) 2012.

Figure 6 shows a time series of the number of unique visitors daily from 2006 to 2012. Typhoons with a high number of unique visitors are also denoted. The system was stopped for inspection in the blank period from Sept. 18 to Sept. 20, 2010 and from Oct. 9 to Oct. 11, 2010. The system was also stopped from Mar. 12 to Mar. 21, 2011, due to the Great East Japan Earthquake.

Figure 7 shows a typhoon track for which a high number of unique visitors were counted. The typhoons with very high number of unique visitors generally passed through the Tokai and Kanto regions, except for typhoon 1112 (TALAS) (Figure 7o), which caused heavy damage in the Kii Peninsula.

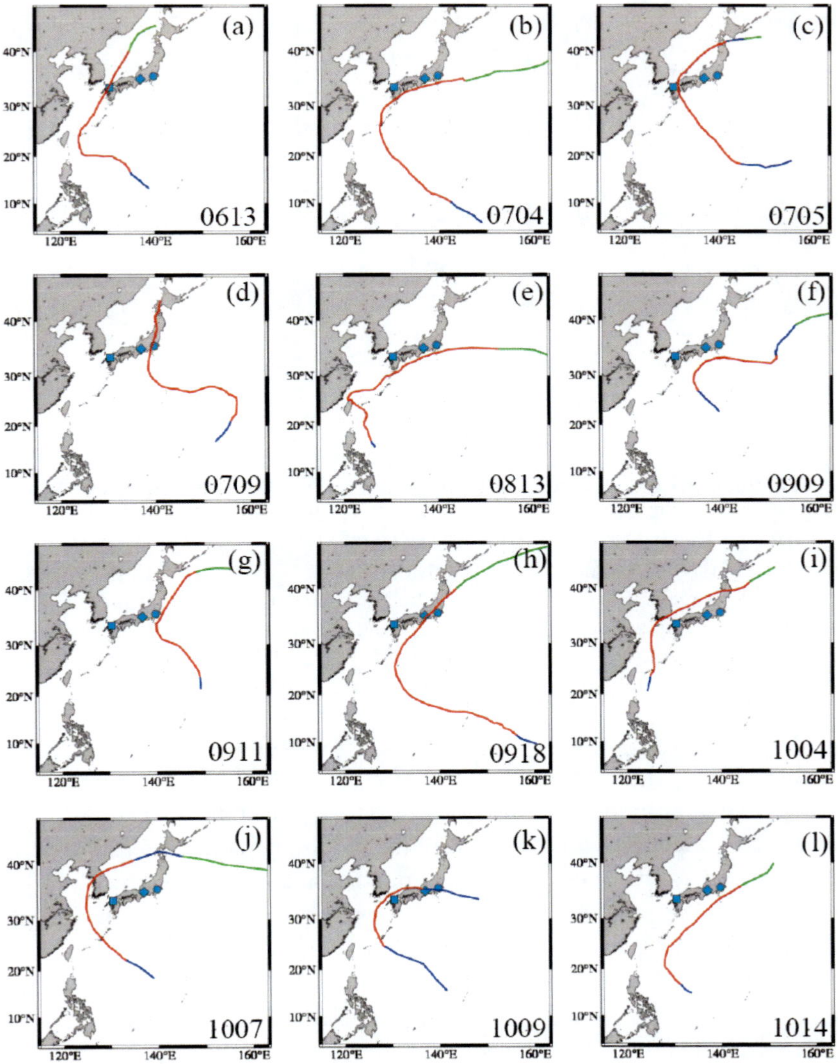

Figure 7. (Continued).

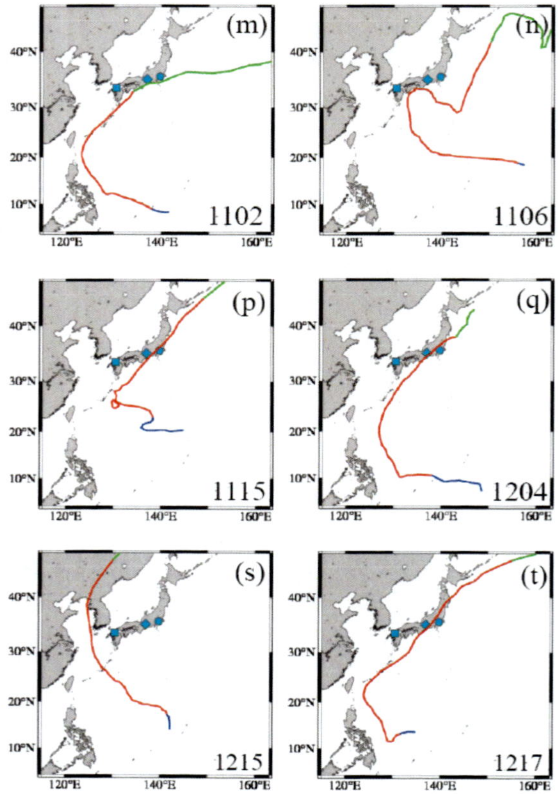

Figure 7. Typhoon tracks with high unique visitors. (a) 0613 (SHANSHAN), (b) 0704 (MAN-YI), (c) 0705 (USAGI), (d) 0709 (FITOW), (e) 0813 (SINLAKU), (f) 0909 (ETAU), (g) 0911 (KROVANH), (h) 0918 (MELOR), (i) 1004 (DIANMU), (j) 1007 (KOMPASU), (k) 1009 (MALOU), (l) 1014 (CHABA), (m) 1102 (SONGDA), (n) 1106 (MA-ON), (o) 1112 (TALAS), (p) 1115 (ROKE), (q) 1204 (GUCHOL), (r) 1210 (DAMREY), (s) 1215 (BOLAVEN), and (t) 1217 (JELAWAT). The blue, red, and green lines indicate a tropical depression, typhoon, and extratropical cyclone, respectively. The circle marker indicates the Tokyo Metropolitan office, the diamond marker indicates the Aichi prefectural office, and the square marker indicates the Fukuoka prefectural office.

Table 2 shows a list of the typhoon number, peak number of unique visitors for each typhoon, maximum wind speed averaged in a specific day, distance between the averaged location of a typhoon on a specific day and the local government office, such as Tokyo, Aichi, and Fukuoka prefectures, human damage (dead/missing and injured), houses destroyed (ruined and half ruined), and houses inundated (inundation above and below floor level).

The day on which the peak number of unique visitors occurred was generally the day the typhoon made landfall in Japan. In other cases, the day on which the peak number of unique visitors was observed was when the typhoon was approaching Japan. The damage situation was obtained from FDMA reports for typhoons 0613(SHANSHAN), 0704(MAN-YI), 0705(USAGI), 0709(FITOW), 0711(NARI), 0909(ETAU), 0918(MELOR), 1106(MA-ON), 1112(TALAS), 1115(ROKE), and 1204(GUCHOL). For other typhoons, information on damage was obtained from reports of local governments. As such, collecting the damage reports related to typhoons may be imperfect.

Table 2. List of the peak number of unique visitors, average daily maximum wind speed with peak number of unique visitors, distance between average daily location of a typhoon with peak number of unique visitors and Tokyo metropolitan government office, Aichi prefectural office, and Fukuoka prefectural office, human damage (death /missing and injured), houses destroyed (ruined and half ruined), and houses inundated (inundation above and below floor level) in each typhoon

Typhoon number	Peak unique visitors	maximum wind speed (knot)	Distance from Tokyo (km)	Distance from Aichi (km)	Distance from Fukuoka (km)	Human damage	Houses destroyed	Houses inundated
0603	114	95	1733	1504	1124	1	0	175
0604	82	55	1920	1633	1019	0	0	0
0607	118	63	390	193	727	0	1	25
0610	121	50	703	483	608	4	3	11
0612	128	65	1042	1352	2081	0	0	0
0613	179	97	2006	1707	1039	458	673	1366
0704	316	72	1054	747	253	86	52	3413
0705	167	81	896	603	282	30	10	637
0709	351	70	325	311	912	93	37	1610
0711	108	35	1305	1090	846	2	36	0
0712	108	97	2018	1712	1010	2	16	0
0715	99	70	1806	1572	1170	0	0	0
0720	80	36	1210	972	701	1	0	0
0807	110	61	2132	1828	1134	0	0	0
0813	271	55	1149	846	273	2	15	674
0815	184	42	1688	1378	659	3	1	5
0909	263	39	648	404	572	50	1313	5602
0911	332	56	189	430	1140	2	0	0
0918	1050	81	871	591	437	142	98	3637
0920	214	69	2090	1804	1184	0	0	0
1004	188	50	1633	1321	591	4	0	243
1007	195	48	1332	1096	776	5	8	0
1009	433	16	309	244	794	4	40	483
1011	159	51	1695	1425	899	5	0	0
1014	298	92	1426	1149	636	8	1	0
1102	219	42	855	561	424	77	9	144
1105	219	49	1964	1705	1193	2	0	1
1106	231	75	776	477	351	57	1	150
1109	99	84	1558	1265	652	42	0	7
1112	467	50	687	429	521	211	3538	22094
1115	437	75	309	309	886	356	1610	7840
1203	122	63	1017	783	617	0	0	0
1204	510	70	695	451	537	80	2	285
1210	216	63	1858	1546	818	1	0	0
1211	93	31	1025	993	1256	0	0	0
1215	240	73	1545	1238	547	10	51	111
1216	126	109	1719	1477	1056	12	124	1320
1217	414	77	558	355	604	160	191	840
1221	153	75	1445	1202	830	3	0	25

A scatter plot of the daily maximum wind speed versus the peak number of unique visitors is shown in Figure 8. The typhoons with highest number of unique visitors are distributed across a wide range of maximum winds, with no tendency of increasing unique visitors with typhoon strength observed.

Next, we investigate the relationship between the peak number of unique visitors and the distance from the typhoon to three large metropolitan areas (Tokyo Metropolis, Aichi Prefecture, and Fukuoka Prefecture). These three prefectures are open to sea and are densely populated. The location of the local government office was used to represent these prefectures. Figure 9 shows a scatter plot of the average distance between the location of the typhoon track and the three prefectural offices versus the peak number of unique visitors.

From Table 2, the typhoons with the highest number of unique visitors were, in descending order, typhoon 0918 (MELOR), 1204 (GUCHOL), 1112 (TALAS), 1115 (ROKE), 1009 (MALOU), and 1217 (JELAWAT). Typhoon 0918 (MELOR) (Figure 7h) landed in Aichi Prefecture and passed through the Tokai and Kanto regions causing heavy damage. The distance from the Aichi Prefecture and the Tokyo Metropolis is large (Figure 9a and 9b). Because the typhoon passed from the sea east of the Amami islands to the Gunma Prefecture through the day on which the peak number of unique visitors was reported, the averaged location of the typhoon was relatively far from both the Aichi Prefecture and Tokyo Metropolis on the day that the number of unique visitors peaked. Typhoon 1115 (ROKE) (Figure 7 p), 1204 (GUCHOL) (Figure 7q), and 1217 (JELAWAT) (Figure 7t) all followed a similar track to that of 0918 (MELOR). These three typhoons had a similar peak number of unique visitors, which occurred on the day the typhoons made landfall, and were a similar distance from the Aichi Prefecture. Typhoon 1115 (ROKE) was very close to the Tokyo Metropolis, passing from the sea south of the Wakayama Prefecture to the sea south of eastern Hokkaido on the day that the number of unique visitors peaked. The reason for the extremely high peak number of unique visitors for typhoon 0918 (MELOR) is unknown. Typhoon 1112 (TALAS) (Figure 7o) passed from the Shikoku region to the Chugoku region. Neither region is densely populated; yet, high peak numbers of unique visitors were reported. The record-breaking rainfall produced by this typhoon may have caused the high number of unique visitors, since it moved very slowly.

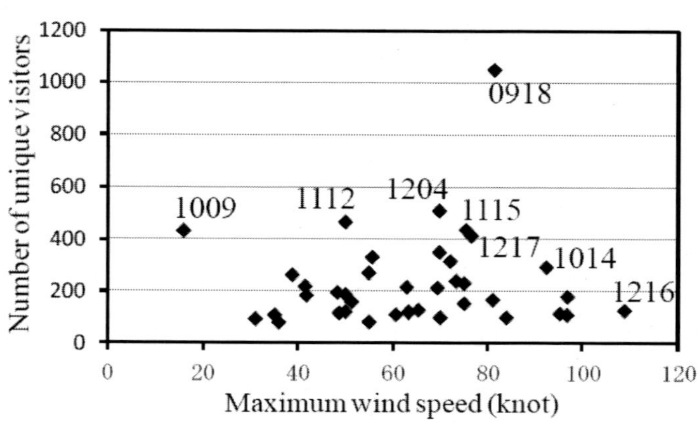

Figure 8. Scatter plot of maximum wind speed versus the peak number of unique visitors for the typhoons in Table 2.

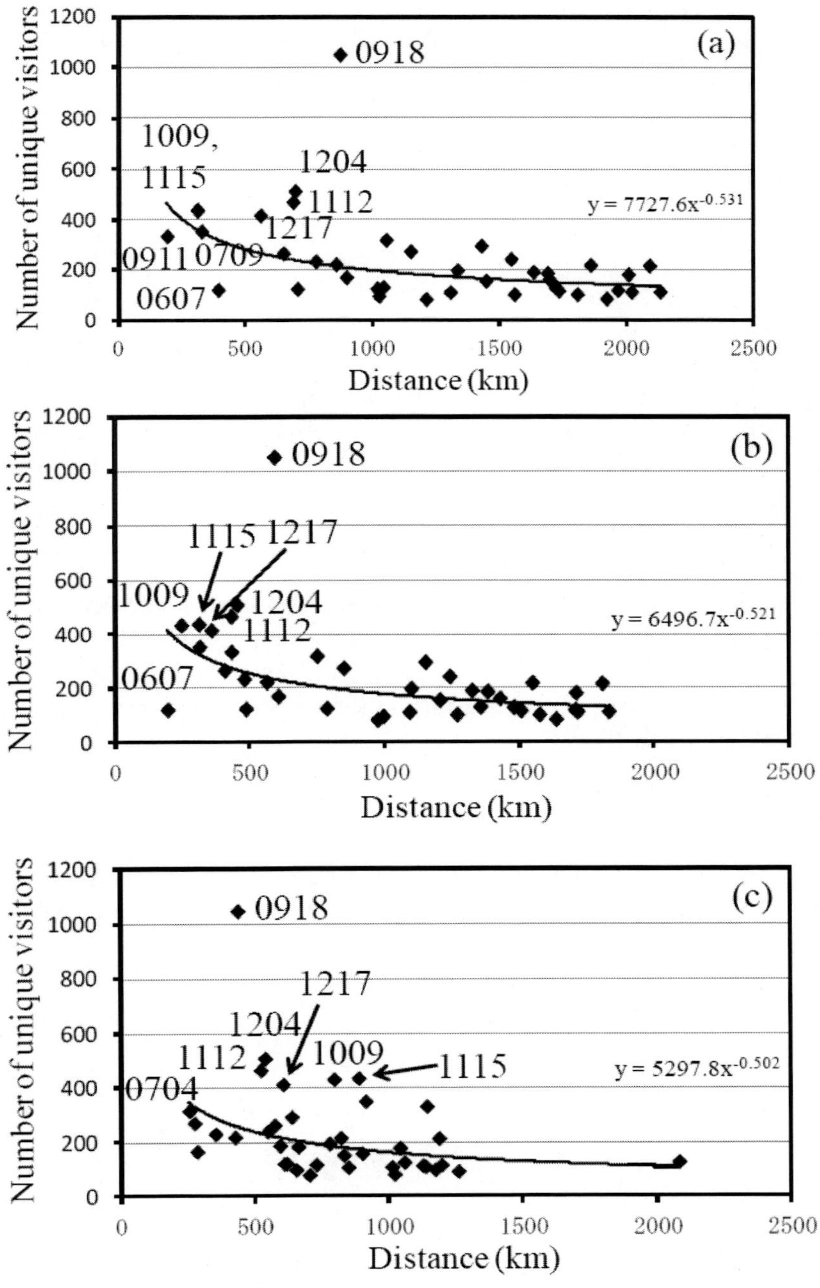

Figure 9. Scatter plot of distance versus the peak number of unique visitors for the typhoons in Table 2. (a) Distance between average daily typhoon track position with peak number of unique visitors and Tokyo Metropolitan office, (b) and (c) same as in (a), but for the Aichi prefectural office and Fukuoka prefectural office, respectively.

The peak number of unique visitors, which was more than 400, is distributed in the upper part of Figure 9c. These typhoons passed through the Tokai and Kanto regions. On the other hand, typhoon 0704 (MAN-YI) (Figure 7b) landed in southern Kyusyu.

The number of unique visitors increased when the typhoon made landfall in the Tokai and Kanto regions, suggesting that the visitors originated from these regions. However, the number of unique visitors does not contain location information in this investigation, and as such, we cannot classify the unique visitors for each region. The location information of users accessing the system is an issue to consider in the future.

CONCLUSION

In this chapter, we introduced the Typhoon disaster database system and discussed the numbers accessing the system. Knowledge of past typhoon disasters is important for reducing forthcoming disasters. In this system, past typhoon disaster information is easily accessible to all. We collected typhoon-related disaster reports since 1951 predominantly from the JMA, FDMA, and local governments in Japan. The damage situation described in the disaster reports was registered in the database. The system has been open to the public via the Internet since July 2005.

The number of unique visitors indicated that when the typhoon was approaching and making landfall, the number of unique visitors increased. The peak number of unique visitors did not depend on the typhoon strength; however, the numbers were high when a typhoon made landfall in the Tokai and Kanto regions, where the population is large.

Considering the increase of unique visitors when a typhoon was approaching and making landfall, it would be useful if the system provided information of past typhoons with tracks similar to that of the approaching typhoon. In the present system, when users point to locations on the map, the system shows past typhoon numbers and their tracks, which passed by the selected points. However, this operation is complicated for the general public; as such, we are currently developing a new page that displays approaching typhoon tracks and past typhoon tracks automatically. Past typhoons are selected when their tracks contain locations similar to the newest location of the approaching typhoon. Thus, users of the new page will be able to see the approaching and past typhoons tracks at a glance. This page is expected to enhance understanding of the past typhoon damage situation for an area, when a typhoon is approaching Japan.

ACKNOWLEDGMENTS

We would like to thank the JMA, FDMA, and local governments for the use of their reports. NCEP Reanalysis data was provided by the NOAA/OAR/ESRL PSD, Boulder, Colorado, USA, from the following web site: http://www.cdc.noaa.gov/.

REFERENCES

[1] WMO, 2002, Proceedings of the Fifth WMO International Workshop on Tropical Cyclones (IWTC-V), http://www.aoml.noaa.gov/hrd/iwtc /IWTC5 _FinalReport.html.

[2] JMA, "Kako no taifu siryo (Past typhoon data)", http://www.data.jma. go.jp/fcd/yoho/typhoon/index.html, In Japanese.

[3] JMA, "Saigai wo motarasita kisyojirei (Weather events that induced disasters)", http://www.data.jma.go.jp/obd/stats/data/bosai/report/index. html, In Japanese.

[4] National Astronomical Observatory of Japan (Ed.), Chronological Scientific Tables, Maruzen Publishing, Tokyo.

[5] M. Yumoto, 2006, Saigai joho, 4, 83-87, In Japanese.

[6] M. Arakida and M. Murata, 2003, Asian Disaster Reduction Center Publication, In Japanese.

[7] JMA, *Geophysical Review*, 620-1238, In Japanese.

[8] E. Kalnay, M. Kanamitsu, R. Kistler, W. Collins, D. Deaven, L. Gandin, M. Iredell, S. Saha, G. White, J. Woollen, Y. Zhu, M. Chelliah, W. Ebisuzaki, W. Higgins, J. Janowiak, K.C. Mo, C. Ropelewski, J. Wang, A. Leetmaa, R. Reynolds, R. Jenne, and D. Joseph, 1996, *Bull. Amer. Meteor. Soc.*, 77, 437-471.

[9] FDMA, "Saigai joho (Disaster report)", http://www.fdma.go.jp/, In Japanese.

[10] S. Shimokawa and T. Kayahara, 2013, *J. Disaster Res.*, 8, 473-483.

In: Natural Disasters
Editor: Karina Linwood
ISBN: 978-1-63463-309-3

Chapter 7

Is It Possible That Rainfall Could Trigger More Large (5.0 ≧ML ≧3.0) Earthquakes? A Case Study for Taiwan

Lin Jyh-Woei *
Dept. of Earth Science, National Cheng Kung University,
Tainan City, Taiwan

Abstract

This study objectively investigates the potential of rainfall that could trigger the earthquake from the earthquake number (5.0 ≧ML ≧3.0) and the information of rain during the period from January 1995 to August 2012. (Lin 2013) reported the similar work that the typhoons could trigger small earthquakes easily especially after Chi-Chi earthquake during January 1995 to July 2012. The earthquakes are likely to have significant positive relationship with rainfall after the Chi-Chi earthquake with Chi-Square test. The result revealed a significant difference between the correlations of daily-accumulated rainfall values and earthquake activities both before and after the Chi-Chi earthquake. The significant difference was discussed regarding) changes of crustal conditions after the Chi-Chi earthquake.

Keywords: Ordinary earthquakes, Chi-Chi earthquake, Earthquake activity, Chi-square test

1. Introduction

The topic of rainfall and typhoon-induced earthquakes has been researched (Husen et al. (2007; Lin 2013). (Husen et al. (2007) reported a series of 47 local earthquakes in the central Switzerland Alps that occurred three days after record rainfall during August 19-23, 2005. In

* Tel: +886-6-2757575 ext 65425. pgjwl1966@gmail.com.

studies conducted in Denver, Colorado, liquid waste was injected into geological faults, resulting in increased seismicity at depths of up to 4 to 5 km. The researcher recorded 1500 minor earthquake events, which subsided after the injections ceased between 1962 and 1966. (Lin 2013) researched the typhoons triggered small earthquakes easily after Chi-Chi earthquake, with 102 typhoons during January 1995 and July 2012.

In this study, a statistical analysis of earthquakes in Taiwan island between January 1995 and August 2012 is performed to test whether rainfall can trigger large earthquakes with the Richter Magnitude (5.0 $\geqq$ML $\geqq$3.0) (Table.1 and Table.2).

The daily average-accumulated rainfall values are the averages of the rain gauge stations near the epicenters of the earthquakes and the earthquake activity located offshore of Taiwan is not considered. The true monthly accumulated rainfall values are the multiple of 28, 30 or 31 (days) for the daily average-accumulated rainfall values (a month period). The true monthly accumulated rainfall values were also used by Chi-square test, the results were the same as using the daily average-accumulated rainfall values, and however the computing time is much more. Thus the daily average-accumulated rainfall values were used due to almost same period of a month (28, 30 or 31 (days) for a month period).

2. Analysis Method

The Chi-square test which is a type of statistical analysis, is conducted to confirm the correlation between two independent data sets (same dimension), and an asymptotic significance value (p-value) is then used to determine significant differences and strong relationships between the two data sets that the p-value is adequately small ($p < 0.05$) (Bewick et al 2004; Lawrence et al. 2006).

Table 1. This table lists the earthquake activity related to daily accumulated rainfall values from January 1995 to September 1999

Year	Month	Earthquake Number	Daily accumulated rainfall values
1995	1	4	22.0mm
1995	2	3	50.0mm
1995	3	2	49.0mm
1995	4	5	70.0mm
1995	5	10	40.0mm
1995	6	4	300.0mm
1995	7	6	70.0mm
1995	8	2	100.5mm
1995	9	1	120.0mm
1995	10	3	20.5mm
1995	11	10	26.5mm
1995	12	6	30.5mm
1996	1	4	10.0mm
1996	2	2	100.0mm
1996	3	5	50.5mm
1996	4	3	48.5mm

Year	Month	Earthquake Number	Daily accumulated rainfall values
1996	5	2	32.8mm
1996	6	1	80.5mm
1996	7	4	90.5mm
1996	8	2	100.5mm
1996	9	4	70.0mm
1996	10	3	40.5mm
1996	11	4	30.4mm
1996	12	3	10.5mm
1997	1	8	60.0mm
1997	2	4	22.5mm
1997	3	4	63.5mm
1997	4	3	68.0mm
1997	5	2	55.0mm
1997	6	6	33.5mm
1997	7	3	35.0mm
1997	8	5	60.0mm
1997	9	3	35.5mm
1997	10	3	40.5mm
1997	11	7	55.5mm
1997	12	3	32.5mm
1998	1	6	34.0mm
1998	2	4	75.0mm
1998	3	3	43.5mm
1998	4	4	41.5mm
1998	5	6	48.5mm
1998	6	3	36.5mm
1998	7	8	47.0mm
1998	8	2	45.0mm
1998	9	5	120.5mm
1998	10	2	100.5mm
1998	11	7	28.5mm
1998	12	5	30.0mm
1999	1	6	20.5mm
1999	2	4	10.0mm
1999	3	2	20.5mm
1999	4	4	70.5mm
1999	5	8	60.5mm
1999	6	3	40.5mm
1999	7	7	12.0mm
1999	8	8	102.5mm
1999	9	90	140.5mm

Source: Central Weather Bureau, Taiwan. Note: The daily accumulated rainfall values are the averages of the rain gauge stations near the epicenters of the earthquakes and the earthquake activity located offshore of Taiwan is not considered.

Table 2. This table shows the earthquake activity related to daily accumulated rainfall values from October 1999 to August 2012 and these data condition is the same as those of Table 1

Year	Month	Earthquake Number	Daily accumulated rainfall value
1999	10	93	40.0mm
1999	11	34	20.0mm
1999	12	17	30.0mm
2000	1	13	45.0mm
2000	2	10	3.0mm
2000	3	17	20.0mm
2000	4	14	20.5mm
2000	5	50	300.5mm
2000	6	59	320.5mm
2000	7	60	434.0mm
2000	8	70	481.5mm
2000	9	51	294.0mm
2000	10	84	500.5mm
2000	11	29	20.5mm
2000	12	40	20.0mm
2001	1	30	100.0mm
2001	2	40	150.0mm
2001	3	25	26.5mm
2001	4	15	28.5mm
2001	5	15	11.8mm
2001	6	60	324.5mm
2001	7	15	12.5mm
2001	8	23	28.5mm
2001	9	33	95.0mm
2001	10	21	15.5mm
2001	11	20	30.4mm
2001	12	23	40.5mm
2002	1	14	3.0mm
2002	2	10	30.0mm
2002	3	36	40.5mm
2002	4	50	408.0mm
2002	5	60	320.0mm
2002	6	40	300.5mm
2002	7	19	45.0mm
2002	8	37	267.0mm
2002	9	27	47.5mm
2002	10	71	300.5mm
2002	11	17	50.5mm
2002	12	19	40.5mm
2003	1	18	24.0mm
2003	2	21	65.0mm
2003	3	16	23.5mm
2003	4	21	340.5mm
2003	5	20	40.5mm
2003	6	38	60.5mm
2003	7	24	30.0mm
2003	8	22	15.0mm
2003	9	24	29.5mm
2003	10	25	22.5mm
2003	11	21	28.5mm
2003	12	80	470.0mm

Year	Month	Earthquake Number	Daily accumulated rainfall value
2004	1	26	44.5mm
2004	2	28	50.0mm
2004	3	21	34.5mm
2004	4	17	110.5mm
2004	5	47	290.5mm
2004	6	13	30.5mm
2004	7	17	120.0mm
2004	8	19	90.0mm
2004	9	16	80.5mm
2004	10	20	40.5mm
2004	11	28	70.5mm
2004	12	9	4.5mm
2005	1	43	70.5mm
2005	2	67	390.0mm
2005	3	50	370.5mm
2005	4	77	389.0mm
2005	5	106	397.5mm
2005	6	52	140.0mm
2005	7	29	133.0mm
2005	8	22	50.5mm
2005	9	20	22.0mm
2005	10	33	4.0mm
2005	11	20	6.0mm
2005	12	21	20.0mm
2006	1	37	43.0mm
2006	2	29	50.0mm
2006	3	24	120.0mm
2006	4	81	400.5mm
2006	5	17	49.0mm
2006	6	36	49.0mm
2006	7	21	33.0mm
2006	8	35	46.0mm
2006	9	19	28.5mm
2006	10	27	43.0mm
2006	11	29	40.0mm
2006	12	47	204.5mm
2007	1	31	200.0mm
2007	2	44	240.5mm
2007	3	50	30.0mm
2007	4	37	200.0mm
2007	5	26	40.0mm
2007	6	40	200.0mm
2007	7	31	220.0mm
2007	8	37	190.0mm
2007	9	40	345.0mm
2007	10	29	240.0mm
2007	11	28	13.0mm
2007	12	28	20.0mm
2008	1	22	30.0mm
2008	2	49	100.0mm
2008	3	32	90.5mm
2008	4	50	254.5mm
2008	5	37	59.0mm
2008	6	39	50.0mm
2008	7	45	250.5mm
2008	8	58	300.5mm

Table 2. (Continued)

Year	Month	Earthquake Number	Daily accumulated rainfall value
2008	9	26	30.0mm
2008	10	29	40.5mm
2008	11	47	100.0mm
2008	12	30	90.5mm
2009	1	49	100.5mm
2009	2	31	80.0mm
2009	3	40	100.0mm
2009	4	27	100.0mm
2009	5	27	50.0mm
2009	6	130	480.0mm
2009	7	119	367.0mm
2009	8	57	400.0mm
2009	9	29	200.5mm
2009	10	62	100.0mm
2009	11	60	80.0mm
2009	12	92	100.0mm
2010	1	26	20.5mm
2010	2	20	46.0mm
2010	3	69	280.0mm
2010	4	48	300.0mm
2010	5	31	180.0mm
2010	6	39	100.5mm
2010	7	40	400.0mm
2010	8	47	270.0mm
2010	9	73	300.0mm
2010	10	93	330.0mm
2010	11	76	200.0mm
2010	12	35	18.5mm
2011	1	38	12.5mm
2011	2	52	300.5mm
2011	3	71	500.0mm
2011	4	53	220.0mm
2011	5	54	40.0mm
2011	6	49	50.0mm
2011	7	91	600.0mm
2011	8	45	200.5mm
2011	9	53	90.0mm
2011	10	46	80.0mm
2011	11	56	120.0mm
2011	12	45	130.0mm
2012	1	107	400.0mm
2012	2	60	280.0mm
2012	3	60	370.0mm
2012	4	39	60.0mm
2012	5	56	50.0mm
2012	6	160	900.0mm
2012	7	49	500.5mm
2012	8	70	100.5mm

3. Data Analysis and Results

The Chi-square test will be used during the previous time-period to determine the difference in the correlation between daily rainfall and earthquake activity before and after the Chi-Chi earthquake. The estimated p-value is 3.7093×10^{-13}, and it is very less than 0.05. Therefore, a strong significant difference exists between the correlations. Figure 1 shows the relationship between the number of the earthquakes in a month (N) and the daily-accumulated rainfalls (R), and the results of the Chi-Square Test. Black circles and black stars indicate the N-R relationships before and after the Chi-Chi earthquake during January 1995 and August 2012, respectively.

The best-fitted equation before the Chi-Chi earthquake for the N-R relationship is as followed (downward black curve):

$$LogN = 1.849 + 0.129LogR \tag{1}$$

The best-fitted equation after the Chi-Chi earthquake for the N-R relationship is as followed (upward black curve):

$$LogN = 2.056 + 0.333LogR \tag{2}$$

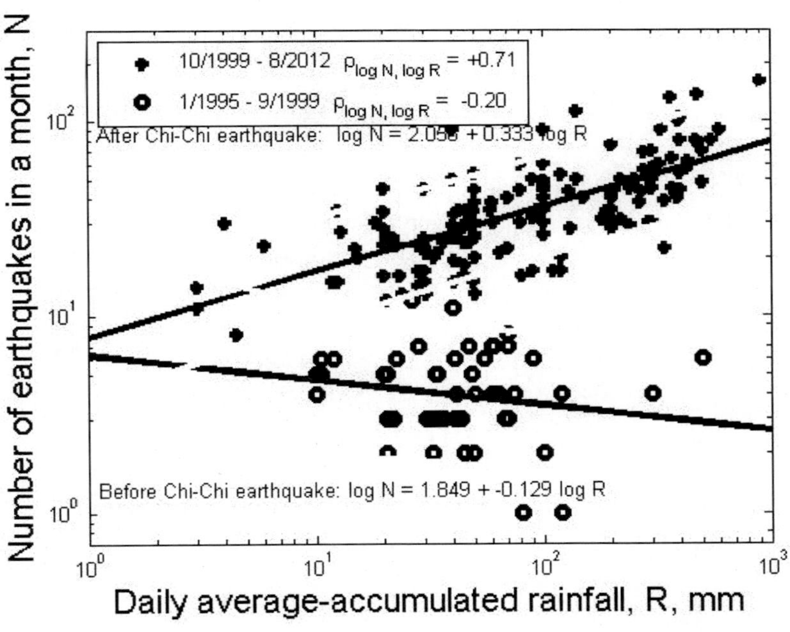

Figure 1. Figure 1 shows the correlations between the number of earthquakes (5.0 ≧ML ≧3.0) in a month (N) and the daily accumulated rainfalls (R) before and after the Chi-Chi earthquake (in log-scale) from January 1995 to August 2012.

The correlation is low, with a slightly negative coefficient of -0.20, before the Chi-Chi earthquake. The N-R relation is approaching the equation (1). Therefore, the earthquake activity has been associated with heavy rainfall during this time period slightly. On the contrary, the correlation is high, with a positive coefficient of +0.71, after the Chi-Chi earthquake, while the N-R relation is approaching the equation (2). During this period, then, the earthquake activity increased during periods of heavy rainfall. These results, covering a period from 1995 to 2012, include activity both before and after the Chi-Chi earthquake that may like create significant changes in crustal stress conditions, allowing rainfalls to trigger small ordinary earthquakes.

4. Discussion

The Chi-Chi earthquake may have resulted in crustal changes in stress conditions throughout Taiwan, forming some blind faults (Tsou et al. 2011) and causing existed faults to become unlocked. Such situation may have then allowed rainfall to trigger earthquakes. Previous results indicate the high likelihood of rainfall-induced earthquakes due to the creation of non-rigid layers after the Chi-Chi earthquake. True cause is arguable; although some possible reasons would be discussed in section one. One potential cause is worth mentioning that level and pressure variances in groundwater due to rainfall (Yeh et al. 2008; Saar et al. 2003). Then water from rainfall may change groundwater levels while simultaneously inducing slight stress and pressure redistributions near the faults according to Mohr-Coulomb failure criterion (Hoek et al. 2002), thereby triggering more large earthquakes although their depths are a little thick.

Conclusion

This study found that the correlation between daily rainfall and earthquake activity, before and after the Chi-Chi earthquake, is significantly different. It is possible that changes caused by the Chi-Chi earthquake made crustal conditions suitable for subsequent rainfall-triggered earthquakes. Potentially, heavy rainfall might have triggered two particular earthquakes observed in this study.

Acknowledgments

The author is grateful to Central Weather Bureau, Taiwan for the data support and in memory of my Wife, Cheng-Ling, Liu (2008-2013).

References

Bewick, V., L. Cheek and J. Ball, (2004), Statistics review 9: One-way analysis of variance. *Crit. Care,* 8(2): 130–136. Doi: 10.1186/cc2836.

Hoek, E.,C. Carranza-Torres and B. Corkum, (2002). Hoek-Brown failure criterion – 2002 Edition. *Proc. NARMS-TAC Conference*, Toronto, 2002, 1, 267-273.

Husen, S., C. Bachmann, and D. Giardini (2007), Locally triggered seismicity in the central Swiss Alps following the large rainfall event of August 2005, *Geophys. J. Int.* 171, 1126–1134.

Lawrence S. Meyers, L. S., G. Gamst and A.J. Guarino (2006), *Applied Multivariate Research*, Sage Publications, Inc., UK.

Lin, J.W, 2013, An empirical correlation between the occurrence of earthquakes and typhoons in Taiwan: a statistical multivariate approach. *Nat. Hazards* 65:605–634. DOI 10.1007/s11069-012-0382-3.

Saar, O., and M. Manga (2003), Seismicity induces by seasonal groundwater recharge at Mt. Hood, Oregon, *Earth Planet. Sci. Lett.*, 214, 605–618.

Tsou, C. Y., Z. Y. Feng, and M. Chigira (2011), Catastrophic landslide induced by Typhoon Morakot, Shiaolin, Taiwan. *Geomorphology,* 127, 166–178.

Yeh, H. F., C. C. Lee and C. H. Lee, (2008), A Rainfall-Infiltration Model for unsaturated Soil Slope Stability, *J. Environ. Eng. Manage*, 18(4), 261-268.

In: Natural Disasters
Editor: Karina Linwood

ISBN: 978-1-63463-309-3

Chapter 8

MODELING AND MONITORING FOR LANDSLIDE PROCESSES

V. B. Svalova*
IEG RAS, Moscow, Russia

ABSTRACT

Landslides process is one of the most widespread and dangerous processes in the urbanized territories. In Moscow the landslides occupy about 3 % of the most valuable territory of the city. There are near 15 places of deep landslides and some hundreds of shallow landslides in Moscow. In Russia many towns are located near rivers on high coastal sides. There are many churches and historical buildings on high costs of Volga River and Moscow River. The organization of monitoring is necessary for maintenance of normal functioning of city infrastructure in a coastal zone and duly realization of effective protective actions. Structure of monitoring system for urban territories is elaborated.

Last years the landslide process activization took place in Moscow.

One of landslide sites is on Vorob'ovy mountains, on a high slope of the right coast of the river Moscow with height of 65 m. In June 2007 rather big landslide took place there near ski-jump.

Another landslide site is in a southeast part of Moscow, occupying the right coast of river Moscow near museum - reserve "Kolomenskoye". Last serious activization of a landslide has taken place there in 2002 with a motion on 53 cm.

Catastrophic activization of the deep blockglide landslide in the area of Khoroshevo in Moscow took place in 2006-2007.

Deep landslide activization took place in Moscvorech'e, Moscow, in September 2009.

Such complicated situation demands development of new investigations of landslide prone zones.

Another part of activity was connected with Russia-Taiwan project "Theory and Methods of Earthquake Early Warning Systems for Underground Pipelines and Hazardous Slopes". Seismic monitoring post was established in Chelungpu fault area near Chiufengershan landslide for seismological phone measurements.

* Email: inter@geoenv.ru

Analysis of seismological phone changes will give possibility to elaborate early warning system.

One of research methods for landslide processes is mechanical-mathematical modeling of gravitational movement of matter on landslide slope. At different stages of the process the development of landslide movement can be described by various mechanical and rheological models.

Mechanical-mathematical model of high viscous fluid was used for modeling of matter behavior on landslide slopes. Equation of continuity and an approximated equation of the Navier-Stockes for slow motions in a thin layer were used. The results of modelling give possibility to define the place of highest velocity on landslide surface, which could be the best place for deformation monitoring post position. Model can be used for calibration of monitoring equipment and gives possibility to investigate some fundamental aspects of matter movement on landslide slope.

Keywords: landslide, modeling, monitoring, cultural heritage, Moscow, Russia

INTRODUCTION

The geological environment is complex dynamic system which change occurs in result of slowly and deep developing processes. The mechanism of landslide process is quite often predetermined by structural - tectonic conditions of territory and lithological-petrographic characteristics of geological composition of rocks. Features of the mechanism of various genetic types of landslips in many respects cause the deepness and volumes of deformation of massifs of rocks, character and speeds of their moving at various stages of development of landslides process. Formation of an initial landslip of the certain type is preceded with usually long period of its preparation during which there is a rearrangement of a site of a slope, decomposition of rocks in slope massif, change of key parameters of a field of pressure, structure, a condition and properties of the rocks composing a slope. Zones of active development of explosive and shift infringements, processes of superficial and deep creep are formed. After overcoming peak resistance of rocks to shift progressing destruction in a zone of landslide displacement with transition of process of deformation in a stage of the basic displacement, quite often catastrophic character with big destructive energy begins. System of environmental monitoring is needed to forecast and prevent such events.

1. MONITORING ORGANIZATION FOR URBAN TERRITORIES

Landslides process is one of the most widespread and dangerous processes in the urbanized territories. In Russia many towns are located near rivers on high coastal sides. There are many churches and historical buildings on high costs of Volga River (Figures 1-8).

In Moscow with landslips it is struck about 3 % of the most valuable territory of city (Figure 9).

First, this territory has value from the historical point of view as the majority of ancient settlements concentrated on coast of the river Moscow and its inflows. In particular, in a southeast part of Moscow in landslides zone there was a museum - reserve “Kolomenskoye” with church and manor constructions of XVI-XVII centuries. In the central part of Moscow, opposite to sports complex "Luzhniky" on landslide slope of Vorob'evy mountains it is located Andreevsky monastery (XVII century). Second, coast of the river Moscow and its inflows are built up in many places by modern multi-store buildings of various purpose, occupied with the industrial, scientific and sports - cultural complexes supposing a mass congestion of people on or close to landslides slopes. In zones of landslides influences there is a number of bridge transitions, numerous communications of city (Kutepov et al., 2002, 2004).

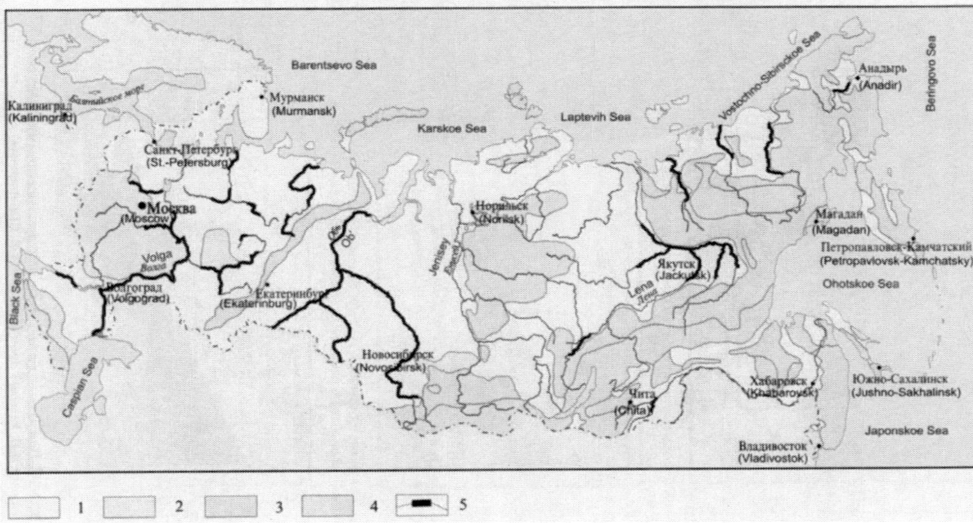

Figure 1. Landslides in Russia (Mjagkov S.M. in Osipov et al. 2002). Plain areas: 1 – few and seldom landslides; 2 – many and often landslides. Mountain areas: 3 – often landslides; 4 – possible very big seismic landslides. 5 - landslides of river valleys.

Figure 2. Church and settlement on the shore of Volga River at Mishkin Town.

Figure 3. Town Uglich. Kremlin.

Figure 4. Town Ples: View of the Kholodnaya Hill. In the forefront, houses of merchants Groshev and Podgorny where I.I. Levitan's studio was located in 1889. To the right: St. Barbara Church.

Figure 5. Town Ples: The summer Church of the Trinity.

Figure 6. Town Ples: View of Zarechye and the Shokhonka river.

Figure 7. Town Tutayev: Resurrection cathedral built in 17 century.

Figure 8. Town Tutayev: Kazan Mother of God church on the left high landslide prone shore of Volga River.

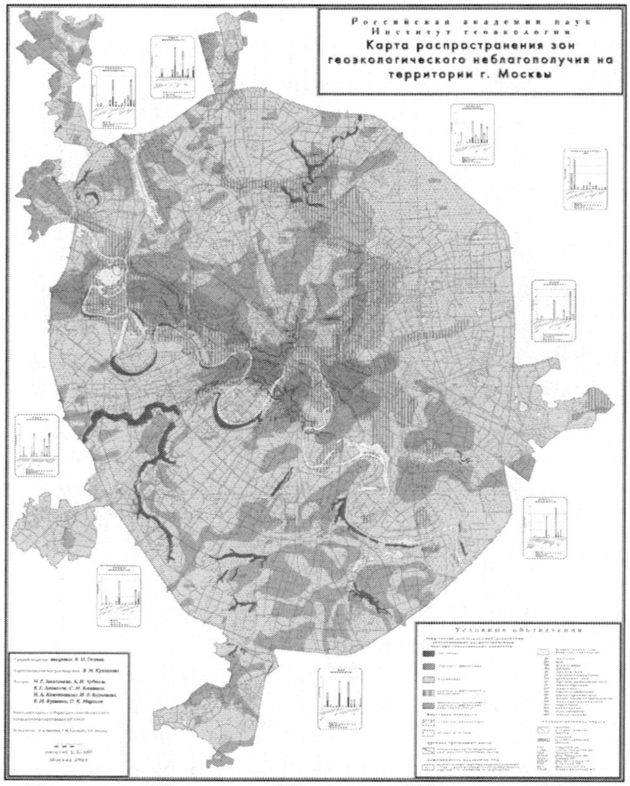

Figure 9. Map of geological hazard zones in Moscow (landslides in red). (IEG RAS, Osipov et al. 2001).

The organization of monitoring is necessary for maintenance of normal functioning of city infrastructure in a coastal zone and duly realization of effective protective actions at a modern level of system of the automated monitoring of landslides process in Moscow according to the general concept of ecological monitoring of underground waters and dangerous geological processes of territory of Moscow.

Stationary supervision over landslips in territory of Moscow are conducted since 1954. For expired time the extensive material on conditions of development and dynamics of landslips is saved up. Value represents not only experience of realization of regime supervision for landslides process and the factors determining its mode, but also the revealed laws of behavior of such complex landslips, as deep block landslips of compression and landslips of expression, similarly to a site of Vorob'evy Mountain.

Last years the regime supervision spent on landslides sites, are considerably reduced, besides the used complex of supervision does not correspond to modern requirements on structure and a level of automation.

On the basis of features of development and a mode of landslips of various types the system of monitoring most full adequate to problems of monitoring of landslides process and requirements to an observant network is offered.

The purpose of monitoring of landslides processes of territory of Moscow are a supply with information of authorities, managements of an ecological condition of the geological environment of city about a condition and dangerous development of landslips for good safety the population and normal functioning of urban services.

Primary goals of monitoring are the following.

1. Revealing sites of active development of landslides process.

Activity is estimated on: to sizes of landslides deformations determined by tool methods of supervision; to results of inspection with mapping of fresh landslides cracks, water-displays; to geophysical parameters of development of landslides motions.

2. The control, including automated, of conditions of landslides slopes and landslides – prone sites.

In Moscow it is revealed 18 sites with display deep block landslips representing the greatest danger to a city infrastructure. Means of monitoring should allow to identify a condition of local sites and a slope as a whole (changes of controllable parameters at a stage of preparation of motions, local displays of integument landslips, deep motions etc.).

3. Revealing laws and the mechanism of development of landslides process and, in particular, deep block landslips.

On the basis of the analysis of the given regime supervision capacity of a landslip, time and conditions of occurrence deep landslides motions, deformation behavior of various parts of landslides slope, interrelation between landslides-forming factors and landslides sites, influence of man-caused factors is established and specified.

4. The prevention of possible influence of landslides deformations on constructions and the communications of city located in landslides zones.

By results of the automated monitoring and regime supervision operative estimations and forecasts of a condition of landslides slopes on sites of an arrangement of various constructions are made. The appropriate preventions are dispatched under the established rules.

5. Gathering, processing, a data storage on controllable parameters and sites, use given others monitoring systems (private, departmental etc.).
6. An estimation and forecasting of a dangerous condition of landslides slope. Calculations of stability and a condition of landslides slope in view of conditions of development of landslides process by the current moment and the required period of anticipation.

On the basis of the given supervision the site of a surface of sliding, a level of underground waters is specified, engineering - geological cross-sections on observably ranges are under construction. Calculations of stability of landslides slope and landslide pressure are made in view of conditions of development of landslide process by the current moment and possible changes of values of influencing factors. By results of change of the controllable parameters determining a condition of landslides slope, the forecast of activity of landslips in Moscow is done.

7. Development of recommendations for acceptance of managing decisions, to increase of stability and stabilization of landslides slopes, to optimum development of slope territories.

During conducting monitoring, accumulation of the information on activity of landslides process, the criteria determining a degree of activity and activization of landslips, dangers for existing in landslides zone of constructions are developed. Recommendations, technical decisions on stabilization of landslides slopes are developed in view of the information received by results of monitoring on features of the mechanism and dynamics of development of landslides deformations, change of risk of construction of various objects in landslides zone.

8. An estimation and the control of efficiency of against-landslides actions and a condition of armed landslides slope.

Means of monitoring of landslides deformations, stress-deformed state of landslides slopes, landslides-forming factors, supervise change of a condition of landslides slope and its parts, characterizing a degree of activity of landslides process and efficiency of carried out stabilizing works.

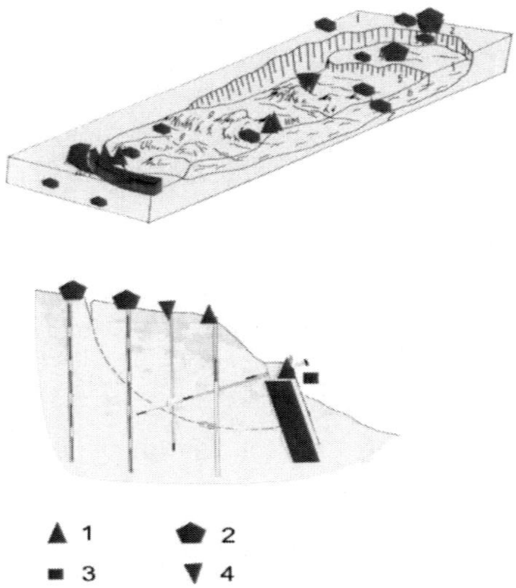

Figure 10. Monitoring post structure for landslide process. (Osipov et al. 2002). 1 – inclinometer, 2 – acoustic seismometer, 3 – deformometer, 4 – water-level meter.

Scheme of structure of monitoring post for landslide slope is performed in Figure 10. At present the project of monitoring network for observation over the landslide slopes in the Moscow River valley is being developed. This network includes the arrangement of the automatic observation system at two slopes (Vorob'evy Hills and Kolomenskoye settlement) and the performance of recurrent geodetic survey at other slopes. The following observations will be performed at the landslide slope in Kolomenskoye region and in Vorob'evy Hills:

- the control over the rock massif displacement by the extensometer data and geodetic reference marks;
- the control of the groundwater level according to special observations;
- the visual description of the slope conditions.

Monitoring system of Moscow can be recommended for other Russian towns especially along rivers.

Results

Monitoring system for Moscow is elaborated. The most dangerous places are connected with landslide prone slops of Moscow River. The experience of monitoring organization can be used in other towns and urban territories in Russia and abroad.

2. Landslide Processes Activization in Moscow

Landslide motion is extremely actual and difficult problem which decision is necessary for preservation of valuable historical monuments and modern city constructions. There are near 15 places of deep landslides and many shallow landslides in Moscow (Figure 11).

Figure 11. Landslides danger places in Moscow (in red).

Figure 12. Vorob'yovy mountains. Landslide slope.

Last years the landslide process activization took place in Moscow.

One of landslide sites is on Vorob’yovy mountains, on a high slope of the right coast of the river Moscow . Within the limits of a considered site there is a historical monument of federal value – (Andreevsky monastery), based in 1648. (Figure 12,13).

Also there are the complex of buildings of Presidium of the Russian Academy of Sciences (Figure 14), bridge with station of underground “Vorob’yovy mountains” and a sports complex . Landslide slope (Figure 15) is in an active condition. In June 2007 a rather big landslide took place there near ski-jump (Figure 16).

Figure. 13. Andreevsky monastery and new buildings on landslide slope.

Figure 14. Presidium of RAS, Andreevsky monastery and new buildings on landslide slope.

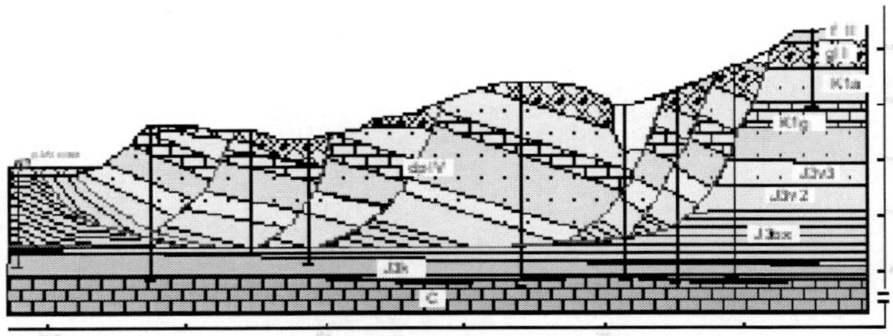

Figure. 15. Schematic cross-section of landslide slope at Vorob'yovy mountains.

Figure 16. Place of landslide activization near ski jump.

Figure 17. Museum - reserve “Kolomenskoye”. Common view . Landslide slope.

Another landslide site is in a southeast part of Moscow near museum - reserve “Kolomenskoye”(Figure 17, 18, 19, 20). Last serious activization of a landslide has taken place there in 2002.

Figure 18. Museum - reserve “Kolomenskoye”. The church of Beheading of the Honest Head of Iowan Predecessor.

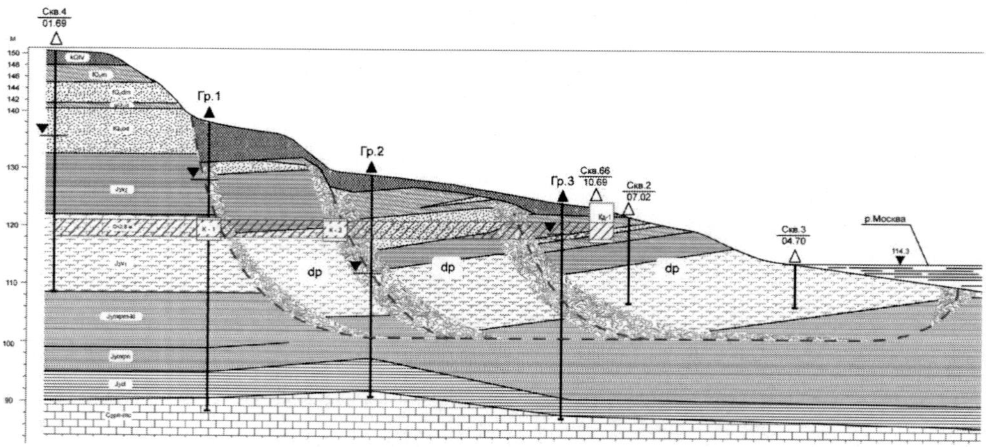

Figure.19. Deep blockglide landslide. Moscow, Kolomenskoye. N.1, N.2, N.3 - extensometers, inclinometers.

Catastrophic activization of the deep blockglide landslide in the area of Khoroshevo on the left-hand shore of the Moskva River at Karamyshevskaya embankment took place in 2006-2007.

Deep landslide activization was in Moskvorechye area in September 2009.

Although landslide monitoring is provided in Moscow from 1930, such a complicated situation demands development of new investigations of landslide prone zones. (Osipov et al. 2002, Kutepov et al. 2002, Postoev & Svalova 2005, Svalova & Postoev 2008).

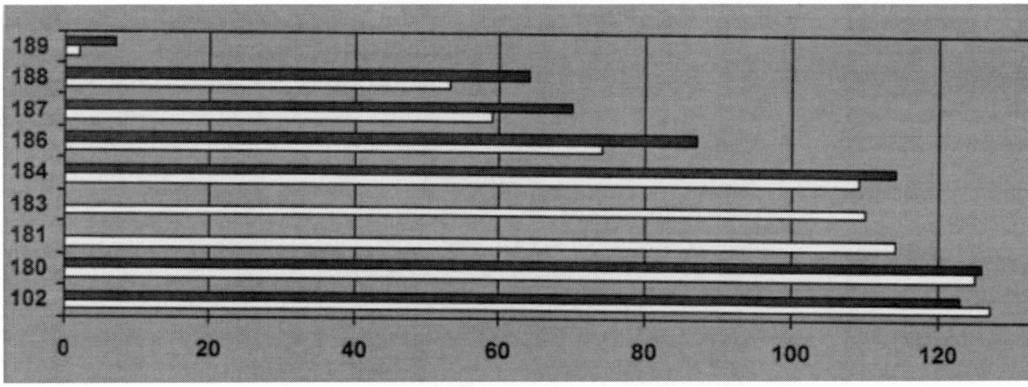

Figure 20. Horizontal displacements (in mm) along landslide movements in 2004 (white), 2005 (black) for bench marks (№№102, 180-189). (Data of Geocenter Moscow).

3. Russia-Taiwan Project "Theory and Methods of Earthquake Early Warning Systems for Underground Pipelines and Hazardous Slopes"

High frequency of landslide in Taiwan area is affected by its geography and geology background. Taiwan is located in the subtropical area with high precipitation, especially during typhoon season in the summer. The surface geometries are highly fractured due to the tectonic activity of collision of Philippine sea plate and Eurasian plate. Landslides are frequently induced by heavy rains or earthquakes in Taiwan Island. Landslide hazard was significantly affected by the 1999 Chi-Chi earthquake, as this event changed the geomorphology and geomaterial conditions.

There are three major components for sustainable management of landslide hazard in Taiwan area: warning system, mitigation plan, and a sustainable urban development plan. The idea of warning system was initiated after the serious debris flow hazard induced by Typhoon Herb in 1996. Various monitoring systems coupled with in-time warning systems were established for large landslide areas, such as Li-shan and Chiufengershan landslide areas. Mitigation plans were implemented and practiced by the central and local governments, which provide unified standards and procedures for hazard mitigation.

A sustainable urban development plan is essential for the landslide hazard management. However, it should consider political and economical factors except technical factors. Besides, the climate change impact to natural hazard becomes more and more significant in Taiwan area. The rainfalls are more concentrated and the temperature increased about 1.5 degrees. The concentrated rainfalls might more significantly affect the landslides in the futures.

Figure 21. Landslide in Taichung area, Taiwan.

Figure 22. Chiufengershan landslide in Chelungpu fault area, Taiwan.

Russia-Taiwan joint research were provided in 2008-2011. Landslide zones in Russia and Taiwan were investigated (Figure 21-26) (Keh-Jian Shou et al.2009, Nikolaev et al. 2011 , Svalova 2009, 2011, 2012, 2013, Zubko et al. 2009). Geophysical devices were elaborated and installed (Figure 25, 26). Seismometer SM-3KV.1 (Russia) was designed for transformation of vertical or horizontal components of seismic oscillations vector. Seismometer enables to use it for high-resolution systems of seismic monitoring, for mesurements of intensity variations and spectral structure of micro seismic background, detection of earthquakes and explosions, measurement of oscillatory movements of constructions, dwellings, etc. The seismometer does not need any adjustment after installation at the workplace. There is no necessity to open the hermetical case of the device. It is checked up by the metrological control.

Figure 23. Russia-Taiwan research group.

Figure 24. Geophysical station near Chiufengershan landslide.

The main purposes of seismic monitoring were:

1. Measurement of variations of seismic activity of region
2. Use of seismic data for emergency switch off the systems at excess of threshold value.
3. Use of information about P - wave arrival from powerful earthquake before destructive S - wave arrival for alarm system construction
4. Forecasting the movements of landslides, as background sound level increases during initial stage of landslide moving as the result of solid crippling of landslide matter.

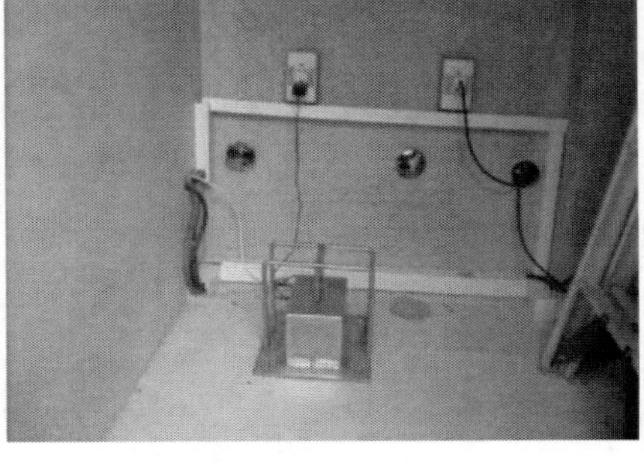

Figure 25. Seismometer CV-373 (Taiwan-Japan).

Figure 26. Seismometer SM-3KV.1. (Nikolaev, Bashilov, Zubko et al., Russia).

Conclusions

In Taiwan Island the landslides are frequently induced by heavy rains or earthquakes. There are three major components for sustainable management of landslide hazard in Taiwan area: warning system, mitigation plan, and a sustainable urban development plan.

4. Mechanical-Mathematical Model for Landslide Processes

Field and theoretical researches for landslides were provided in Russia (Moscow, Volga-river area, Sochi, Caucasus in frame of the First Russian-Indian Summer School on High Performance Computing and their applications to problems in geophysics 2013), in Taiwan on the base of Russia-Taiwan Project "Theory and Methods of Earthquake Early Warning Systems for Underground Pipelines and Hazardous Slopes", in Italy on the base of Russia-Italy project "Risk evaluation and monitoring of natural hazards for urban territories in Italy and Russia", in India on the base of Russia-India project "Mitigation of Natural and Human Induced Disasters" (Keh-Jian et al. 2009; Nikolaev et al. 2011; Postoev & Svalova 2005; Svalova & Postoev 2008; Zubko et al. 2009).

One of the methods of studying of landslide processes is mechanical-mathematical modelling of gravitational movement of the matter on landslide slope. At different stages of the process development the landslide movement can be described by various mechanical and rheological models. At the stage of formation of cracks, losses of stability, break of blocks the models of the elastic medium and model of destruction are applied. During slow movement of soil on the slope the model of high viscous incompressible fluid can be used. Such model allows to calculate velocities of movement in the layer of the matter and to compare them with results of velocity monitoring. Boundary conditions of the problem also depend on concrete situation. So, the condition of sticking is used on the bottom border of the layer in case of slow movement. The condition of sliding or more complex boundary condition is possible on the bottom border, if the process of debris flow, underwater landslip or snow avalanche is considered. The choice of adequate model of the process and statement of initial and boundary conditions are the special mechanical problems.

Our approach is rather close to works of Fathani & Nakamura 2005; Lang & Nakamura 1998; Nakamura & Fathani 2002; Nakamura et al. 1989; Suzuki 2001; Tsurugaya 2001. Let's consider movement of landslide masses on the slope as movement of high viscous incompressible fluid described by equation of Navier-Stockes and continuity:

$$\frac{d\vec{v}}{dt} = \overline{F} - \frac{1}{\rho} grad p + \frac{\mu}{\rho} \Delta \vec{v}$$

$$\text{div } \vec{v} = 0 \quad (1)$$

$\vec{v}$ - vector of velocity, F – force of gravity, p – pressure, ρ - density, μ - viscosity, t - time.

Such a model can be successfully used for the lithosphere movements simulation (Svalova 1975) and for soil movements on the slope (Fathani & Nakamura 2005). The problem is to estimate effective viscosity for real matter. It can be done by comparison of calculated and measured velocities of the matter movements. Let the characteristic horizontal scale of a body of landslip L considerably surpasses its thickness h. We shall count also a landslip extended enough in the plan that allows to consider three-dimensional model as two-dimensional one for sections of landslide bodies. Following works (Svalova 1975, 1992, 1993, 2009, 2011, 2012, 2013; Svalova & Sharkov 1992) and applying the method of decomposition on small parameter, it is possible to get the equation of continuity and an approximated equation of the Navier-Stockes in dimensionless form for slow motions in a thin layer (2):

$$\begin{cases} \dfrac{\partial P}{\partial X} = \alpha\mu \dfrac{\partial^2 U}{\partial Z^2} \\ \dfrac{\partial P}{\partial Z} = -\rho \end{cases}$$

$$\frac{\partial U}{\partial X} + \frac{\partial W}{\partial Z} = 0$$

$$\alpha = \frac{F}{R\left(\dfrac{h}{L}\right)^3} \quad (2)$$

$$R = \frac{u_0 L \rho_0}{\mu_0}$$

$$F = \frac{u_0^2}{gL}$$

P is dimensionless pressure, U,W – dimensionless velocities, F - Frude number, R – Reynolds number, ρ - density, μ - viscosity, ρ_0, μ_0, u_0 - scales of density, viscosity and velocity.

Then it is possible to get the velocities and pressure in the layer (3):

$$P = \rho(\varsigma^* - Z)$$

$$U = U_0 + \frac{\rho}{2\alpha\mu}\frac{\partial \varsigma^*}{\partial X}[(\varsigma^* - Z)^2 - (\varsigma^* - \varsigma_0)^2] \tag{3}$$

$$W = W_0 + \frac{\partial U_0}{\partial X}(\varsigma_0 - Z) +$$

$$\frac{\rho}{\alpha\mu}\frac{\partial^2 \varsigma^*}{\partial X^2}\left[\frac{1}{6}(\varsigma^* - Z)^3 + \frac{1}{3}(\varsigma^* - \varsigma_0)^3 -\right.$$

$$\left.\frac{1}{2}(\varsigma^* - Z)(\varsigma^* - \varsigma_0)^2\right] +$$

$$\frac{\rho}{2\alpha\mu}\left(\frac{\partial \varsigma^*}{\partial X}\right)^2 (Z - \varsigma_0)^2 + \frac{\rho}{\alpha\mu}\frac{\partial \varsigma^*}{\partial X}\frac{\partial \varsigma_0}{\partial X}\varsigma_0(\varsigma^* - \varsigma_0)$$

ς_0 - the bottom border of the layer,

ς^* - the top border.

Let on the bottom border the condition of sticking is satisfied (4):

$$U_0 = W_0 = 0 \tag{4}$$

The discharge of matter along the layer is (5):

$$Q = \int_{\varsigma_0}^{\varsigma^*} U dZ = -\frac{\rho}{3\alpha\mu}\frac{\partial \varsigma^*}{\partial X}(\varsigma^* - \varsigma_0)^3 \tag{5}$$

Since Q=const lengthways X, then (6):

$$\frac{\partial Q}{\partial X} = 0$$

$$\frac{\partial^2 \varsigma^*}{\partial X^2}(\varsigma^* - \varsigma_0)^7 + \frac{9\alpha\mu Q}{\rho}\left[\frac{3\alpha\mu Q}{\rho} + (\varsigma^* - \varsigma_0)^3\frac{\partial \varsigma_0}{\partial X}\right] = 0 \tag{6}$$

The condition of convexity of upper boundary is (7):

$$\frac{\partial^2 \varsigma^*}{\partial X^2} < 0 \Rightarrow \tag{7}$$

$$\frac{3\alpha\mu Q}{\rho} > -(\varsigma^* - \varsigma_0)^3 \frac{\partial \varsigma_0}{\partial X} \tag{8}$$

This condition (8) enables to analyze the form of the surface of moving matter (Figure 27).

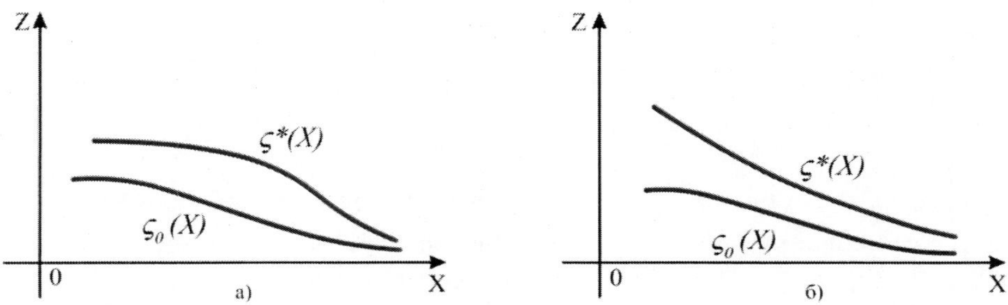

Figure 27. Sketches of dimensionless landslide surface. Various possible forms of landslide ground surfaces: a) - convex, b) - concave.

Structure of clinoforms (convex) can arise, if:

1. Q is large, that is flux is high
2. μ is large. It means that matter spreads bad and can support big angle
3. ρ is small. It means that matter has large specific volume and is friable
4. grad ς_0 is small, that is angle of lower boundary is small
5. $(\varsigma^* - \varsigma_0)$ is small, that is thickness of sedimentary layer is small. Under fixed Q it means that velocity of flux is high and formation of clinoformes and even overturning of rocks are possible

All these conditions seem to be natural enough for explanation of formation of structures such as inflows and clinoforms of sedimentary cover that confirms correctness of the model. It is important to define the place of maximum velocity on the slope. An optimum place for location of monitoring post is the point of maximal speeds of movement of landslide masses.

Let's consider the massif of sedimentary rocks with the top border ς^* representing landslide slope. The bottom border ς_0 is compatible with an axis X. The maximum of horizontal speed U is reached on the top border ς^* of the massif according to condition (9):

$$\frac{\partial U}{\partial Z} = -\frac{\rho}{\alpha\mu}\frac{\partial \varsigma^*}{\partial X}(\varsigma^* - Z) = 0 \Rightarrow Z = \varsigma^* \tag{9}$$

Point of the maximal horizontal speed on the surface ς^* can be found from condition of equality to zero of the first derivative (10):

$$\frac{\partial U^*}{\partial X} = 0, where\, U^* = -\frac{\rho}{2\alpha\mu}\frac{\partial \varsigma^*}{\partial X}(\varsigma^*)^2 \tag{10}$$

From here it is easy to receive the condition (11):

$$\frac{\partial^2 \varsigma^*}{\partial X^2}\varsigma^* + 2\left(\frac{\partial \varsigma^*}{\partial X}\right)^2 = 0 \tag{11}$$

It is necessary to stress, that ς^* (X) is a known function - the surface of landslide slope. And the received condition allows to find a point on a slope where speed of movement is maximal. Let's consider for illustration of the received decision the surface of landslide as (12) (Figure 28):

$$\varsigma^*(X) = -thX + 1 \tag{12}$$

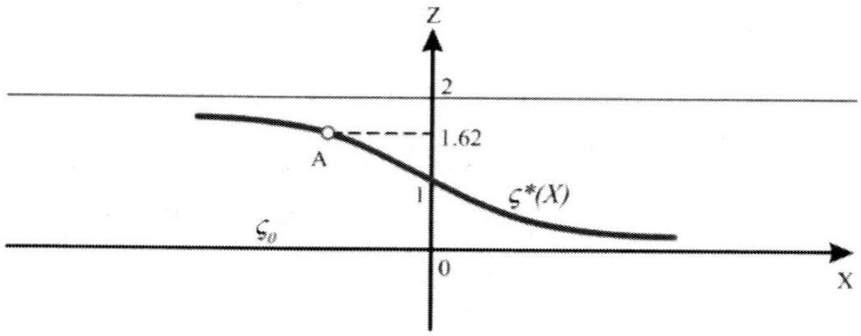

Figure 28. Sketch of dimensionless slope surface. Point A is the point of maximal horizontal speed of movement of masses on the ground surface of the slope.

It reflexes roughly the form of landslide surface. Then the condition (11) gives (13):

$$th^2X - thX - 1 = 0 \tag{13}$$

When we receive (14) $thX = \frac{1-\sqrt{5}}{2}$ and

$$\varsigma^* = \frac{1+\sqrt{5}}{2} \approx 1{,}62 \qquad (14)$$

Such position of the point of the maximal horizontal speed seems to be real, and more exact data on the structure of landslide and its surface will enable to define such point on a concrete slope. The point of maximum speed on a slope defines the place of possible failure of a landslip in case of achievement of limiting pressure in massif of rocks.

There could be several points of local maximum of speed on a slope that characterizes an opportunity of failure of a landslip on each terrace of a slope.

The places of minimum speed can be used for building constructions and oil-gas pipelines. Model velocities of matter can be compared with real velocities that give possibility to define the real mechanical parameters of media and to calibrate the model.

CONCLUSION

Mechanical-mathematical model of high viscous fluid is elaborated for simulation of slow matter movement on landslide slopes. The results of modelling give possibility to define the place of highest velocity on landslide surface, which could be the best place for monitoring post position. Model can be used for comparison of calculated and measured velocities and estimation of effective viscosity of real matter that gives opportunity to calibrate the model. The results of modelling give possibility to investigate some fundamental aspects of landslide matter movement.

REFERENCES

Fathani TF & Nakamura, H (2005) Numerical analysis of the movement distance and velocity of landslides. Proceedings of the International Conference on Landslides, Vancouver, Canada, 10p.

Keh-Jian Shou, Nikolaev A.V., Bashilov IP, Svalova VB, Lin CC, Song ST (2009) Theory and methods of earthquake early warning systems for underground pipelines and hazardous slopes, Abstracts of International Conference Geohazards 2009, Taiwan.

Kutepov VM, Sheko AI, Anisimova NG, Burova VN, Victorov AS et al. (2002) Natural hazards in Russia. Exogenous geological hazards. Moscow, "KRUK", 345 pp.

Kutepov VM, Postoev GP, Svalova VB (2004) Landslide hazards estimation on sites of modern and historical constructions in Moscow. *Proceedings of 32 IGC*, Italy, Florence

Lang YH & Nakamura, H (1998) Characteristics of slip surface of loess landslides and their hazard area prediction. *Journal of Japan Landslide* Society, 35(1), 9-18.

Nakamura H & Fathani TF (2002) Hazard area prediction for landslide debris. *Proceedings of the 10th International Conference and Fieldtrip on Landslide*, Krakow, 129-142.

Nakamura H, Tsunaki R & Ishihama S (1989) Simulation model for debris movement of landslides. Proceedings of the Japan-China Symposium on Landslides and Debris Flows, Niigata, Tokyo, 81-86.

Nikolaev AV, Bashilov IP, Keh-Jian Shou, Svalova VB Manukin A B, Zubko YN, Behterev SV, Kazantseva OS, Rebrov VI (2011) Some directions of works on maintenance of geological safety of engineering constructions. Proceedings of ENGEOPRO, Moscow, 7pp.

Osipov VI, Shojgu SK, Vladimirov VA, Vorobjev YuL, Avdod'in VP et al. (2002) Natural hazards in Russia. *Natural hazards and society*. Moscow, "KRUK", 245 pp

Postoev GP, Svalova VB (2005) Landslides risk reduction and monitoring for urban territories in Russia. Proceedings of the First General Assembly of ICL (International Consortium on Landslides), "Landslides: risk analysis and sustainable disaster management", Washington, USA, Springer, pp 297-303.

Suzuki K (2001) Estimation of the travel distance of landslide debris in granite area. *Master Thesis at Tokyo Univ. of Agriculture and Technology*. Tokyo, 32-47.

Svalova VB (1975) The models of the lithosphere movements. Ph.D. Thesis at Moscow State University, 130 p.

Svalova VB (1992) Mechanical-mathematical models of the formation and evolution of sedimentary basins. *Sci.de la Terre, Ser.Inf.*, Nancy,31, 201-208.

Svalova VB (1993) Mechanical-mathematical simulation of geological structures evolution. *Geoinformatics*, v. 4, N3, 153-160.

Svalova VB (2009) Mechanical-mathematical modeling for sedimentary movement and landslide processes. CD Proceedings of the International Association for Mathematical Geosciences Meeting (IAMG 2009), Stanford, California, USA, 15 pp.

Svalova VB (2011) Mechanical-mathematical modeling and monitoring for landslide processes. *Journal of Environmental Science and Engineering*. V 5, N 10, 1282-1287.

Svalova V (2011) Monitoring and modeling of landslide processes. Monitoring. *Science and technology*. №2(7), 19-27. (in Russian).

Svalova VB (2011) Landslide process simulation and monitoring. Proceedings of ENGEOPRO, Moscow, 7pp.

Svalova VB (2012) Mechanical-mathematical modeling and monitoring for landslides. Proceedings of IPL (International Program on Landslides) *Symposium,* UNESCO, Paris, 63-68.

Svalova VB (2013) Risk reduction for landslide hazards. Modeling and monitoring, Proceedings of The International Conference Natural risks: analysis, estimation, mapping, Moscow, MSU, 157-163. (In Russian).

Svalova V, Postoev G (2008) Landslide Process Activization on Sites of Cultural Heritage in Moscow, Russia. Proceedings of the First World Landslide Forum 2008, Tokyo, Japan, 4p.

Svalova VB & Sharkov EV (1992) Mantle diapirs and surface depression in back-arc areas: a rheological model. *Ofiolity,* 17/1/, 165-170.

Tsurugaya K (2001) Study of the occurrence and flow mechanism of the large-scale landslides caused by an earthquake . *Master Thesis at Tokyo Univ. of Agriculture and Technology*. Tokyo, 80-108.

Zubko YN, Nikolaev AV, Bashilov IP, Svalova VB (2009) Autonomous portable seismic receiver with digital registration for seismological studies. Abstracts of International Conference Geohazards 2009, Taiwan.

INDEX

D

E

F

G

H

I

J

O

P

Q

R

S

T

U

V

W

Y

Z